Ortho-Para V

Alternative-Integrated Medicine

Wallace L. Salzman M.D.

This book was printed in the United States of America.
ISBN 978-1-4303-0753-2

To order additional copies of this book, contact:
Ortho-Para Publishing
www.ortho-para.com

Layout Design by Joanne Ranney and Richard Salzman

Dedication

This book has been written during what probably is the last major crisis of my life. It physically started at the beginning of my 8th decade and I am fortunate that although I am hemiplegic on the left side of my body, the paralysis hit my right brain and not my left. I can still talk and create what I consider excellent material, as can be seen by the contents of this book. It is possibly my last attempt to not only explain human behavior but also examine the processes of treating the population of every age.

To accomplish this end, I needed the emotional and physical support of my wonderful children, Helen, the woman of my life, and the staffs of 4 hospitals, Lee Memorial of Fort Meyer, Florida, Rehabilitation Institute of Chicago, Condell Memorial Hospital of Libertyville, Il, the fabulous, but less known, Brentwood North Nursing Home of Deerfield; Il, and The Park, a senior citizen retirement hotel.

For 18 months I've watched small changes taking place in my body. It has been a combination of atrophy of disuse and a struggling attempt at bringing preexistent strong muscles to previous functional strength. It has only been with the courageous and assiduous attempts of my Therapists in the departments of Physical and Occupational Therapy that some function has returned.

I therefore dedicate this volume to the Angels of my life;

Lee Memorial Hospital; Rehab Institute Chicago; Brentwood North

To: Tod Sunde; Brian; Megan Schanding; Michelle Lawson; Madelyn Delacruz; Kim Peres; Roy Lipio; Frank Gureck; Ferdinand Gerira; Laurie Wedell; Condell Memorial Hospital; Romalyn Cariaca; Karen Lloyd; Matthew Rockey; Jeffrey Brown; Audrey Wright; Elaine Kempers; Paul Raab; Usella Barnhill; Rebecca Kroes; Luz Vinlvan; Mike Whittle; Mary C.Luehrs-Masel;

Elaine Margol; Gail Ann Chaduck; Margaret Graca; Bryan Didley; Fortune Fortunato; Julia Hanft; Cecilia Olivera; Maria Lopez; Helen Herrera; Phoebe Herrera; George Peraza; Monica Jerusalem; Mary Disierra; Rosbiz Chenco

Physicians In Charge

Eileen L Schwartz, M.D.

Harvey M. D.

Charles Colodny

There were and are a far greater number of bit players. Nurses, Aids, House keepers, Secretaries, and even well dressed doctors took an interest in my slow recovery.

To all, I pray for their health and thank them profusely for the help, the smiles, and the Touch therapy that kept me in a positive mode most of the time.

Preface

This is the 5th Treatise that has been designed to incorporate the didactic laws discussed in greater detail in Treatises # I, II, III. and IV.

In Treatise # I, we note 9 Cosmic Laws, 59 Laws of Systems and Particles, and the biologic representation of these laws. In discussing the Para and Ortho polarities and their effect on human and societal behavior. we ultimately discover the influence of these polarities on the create/destroy ratio of all that is life and living.

In Treatise #II we note the Embryonal Evolvement of Life, the Embryonal Eras and Ages, and the Sages that made both individual and societal unfolding possible.

In Treatise #III we offer a means of detecting the autonomic polarity of patients and the therapeutic options that can be offered if any polar extremes are registered in the important systems of involvement. It also discusses the evolvement of civilization based on known history.

Our present Treatise (#V) examines the functional and structural nature of illness and the options offered by Traditional and Alternative care healers to the patient population. It is hoped that it will uncover some of the mysteries of tactile, herbal, pharmacologic and magnetic therapy and an integrated means of determining which therapies to use in the acute and chronic illnesses that have thus far been resistant to the magic of the physician and healer.

Beside traditional allopathic medicine, we will touch on the following alternatives, and postulate in what way they might fit into an Integrative medicine that has already become part of the armamentarium of many highly trained physicians and healers throughout the world. They include the following:

1) Surgery
2) Pharmaceuticals-oral, injection
3) Herbs and foods
4) Acupuncture
5) Manipulation; bodywork, immersion in water and heat, exercise, meditation, dance, drumming, massage

6) Bioenergetics- chanting, touch, hands on, visualization
7) Prayer
8) Meditation- sitting and art
9) Physical therapy
10) Chiropractic
11) Manipulative Osteopathy
12) Homeopathy
13) Faith based and Psychotherapeutic
    approaches
14) Biomedicine
15) Bioenergetics
16) Ayurvedic Medicine
17) Chinese Medicine
18) Art therapy
19) Naturopathy
20) Divine healing
21) Self Healing
22) Hypnosis and Imagery
23) Dietary Medicine
24) Aromatherapy
25) Verbal charm
26) Reflexology
27) Osteopaths
28) Curanderismo
29) Biofeedback
30) Mind-Body
31) Electromagnetic
32) Magnetic
33) Chelation therapy
34) Aikido
35) Holistic Medicine
36) Mesmeric Vital Energy
37) Psycho-neuro-immunology
38) Bone setting
39) Hydrotherapy
40) Charismatics
41) Psychic Energy
42) Acupressure
43) Applied Kinesthesiology
44) Cell Therapy
45) Colon therapy
46) Detoxification

47) Environmental Medicine
48) Enzyme therapy
49) Flower therapy
50) Guided Imagery
51) Humor therapy
52) Hyperthermic therapy
53) Hypothermic therapy
54) Juice therapy
55) Light therapy
56) Music therapy
57) Neural therapy
58) Nutritional therapy
59) Orthomolecular Medicine
60) Oxygen therapy
61) Qigong
62) Reconstructive therapy
63) Shiatsu
64) Yoga
65) Alexander techniques
66) Aston Patterning
67) Feldenkrais
68) Hellerwork
69) Myotherapy
70) Rolfing
71) Trager
72) Color therapy
73) Cold laser therapy
74) Aerobic exercises

It is my hope that some and possibly all of the modalities now in use can be made comprehensible to both the professional and the laity, and by so doing a more benevolent relationship can be established between all who represent the healers of this world.

During the 50 years I was involved with the practice of medicine, as a family Practitioner, my allopathic prejudices blinded me to the possibilities that the underlying sub-structure of knowledge that guided my therapeutic decisions, might have some flaws.

Thus, fixed and absolute notions were weakened only when I represented my hospital at a conference in Stockholm, Sweden on the Morality and Ethics in Medicine.

A young, bright Norwegian philosopher called our attention to the fact that there were no representatives to the conference from any third world nations. She looked at our surprised faces and then proceeded to tell us why.

Whereas the physicians of the western, modern nations approached illness with the notion that if a patient is not well, there is something wrong with a part, and if we can discover what part and correct its pathology, the patient will get well. This form of partistic medicine, except for some old psychiatrists who had yet to break down mental illnesses into their neuro-hormonal excesses or deficits, was the position of most of Medicine.

But this was not the paradigm of the healers, shamans and witch doctors of 3rd World Nations. Their approach was to engage their patient in an exploration into system relationships. Their search would dig into the possible parental, marital, family and tribal disturbances, and also the fears related to GOD, the Spirits and the hereafter. If in the exploration of all the patient's relationships a problem was uncovered, the healer would concentrate on healing the dysfunctional system, exorcise bad spirits, perform rituals that were tribally considered powerful, and drinking brews that the patients knew were curative. The average patient would respond, with both the focal and relational symptoms disappearing at the same time.

Modern Medicine was, and is, treating the Parts, knowing that in curing the Parts, the Whole would get well.

Third World Medicine was, and is, treating the Whole, knowing that when treating and curing the Whole, the Parts would get well.

It reminded me of Sir William Osler's statement at the beginning of the 20th century when he said "Ask not what organism is causing the disease, but instead what has happened to the patient to make him susceptible to the organism." It is probably the first introduction to the concept of neuro-humoral immunity and suggests that in understanding the whole, we might be more successful in our efforts to keep the parts well.

My mind, when young, remained closed to the meaning of Osler. It was an interesting idea, but not a strong enough revelation to influence my practice of medicine.

However 12 years ago I retired and began writing my first three treatises, and thus preparing myself to introduce to the scientific community the thoughts that had pursued me for 45 years. During this time my writing consumed me, and the physical and kinetic aspects of my day slowly dissipated, so that I became an obese couch potato. My blood pressure rose and by putting myself on a diuretic, I provoked an increase in my blood uric acid and eventually developed gout. I now had physical reasons for not walking because I now had hot, swollen feet. All the available medications didn't agree with me and I chose not to treat myself. My daughter Beth, who is a physiotherapist, was bothered by that decision and had a suggestion. She had seen some therapeutic miracles with magnetic therapy and insisted that I place magnetic insoles in my shoes and wear them for several weeks. In two weeks my feet were normal, despite the fact I hadn't changed my diuretic. This alternative approach to my painful feet opened me up to the concepts of Alternative medicine. I was now ready to study and see in what way these magnets could possibly have effected my feet.

Let me digress for a moment and go back 53 years, when I was taking my residency in Internal Medicine. I had wonderful experiences at the Monmouth Memorial hospital where I took care of a large County ward of over 100 beds and also shared in the care of a large private care population. Of all the physicians, there was only one who irritated me, a Dr Victor Knapp who constantly kept reminding me of the hazards of the medicines that had been used during the 19th and early part of the 20th century. His advice sounded like a warning that would be closing the door to the successful use of many therapeutic tools I was taught to use. His examples were overwhelmingly correct, but I resisted his attempt at influencing my very positive attitude regarding my powers as a physician. I did not want to hear his tainting of medicine with warnings regarding iatrogenic fiascos.

Sometime, halfway through my residency, I read an article in the American Journal of Medicine by one of the foremost clinicians of Cornell who recommended the I.V. administration of Digoxin for congestive heart failure. He claimed it to be the fastest way to clear their lungs and I enthusiastically gave my very next patient, a beautiful lady in her 70s, the recommended dose, infused very slowly. Within 3 minutes her heart slowed down to 40, then 30 and it finally stopped. I was mortified. I had taken EKGs before, during and after the infusion, and I could only watch helplessly as this lovely lady died before my eyes. There were no coronary care units, and the emergency electrical devices we have today were not yet invented. I was a novice of 23.

With enormous guilt I presented this case to a clinical pathologic conference of 60 clinicians and expected to be castigated for my act. Instead, they all thanked me for the presentation, the EKGs, and shared with me the fact that there are moments of glory in Medicine, and also sad moments when all we do is fruitless and we must get comfortable with the inexorable fact that death is the ultimate consequence of life.

Dr. Knapp was sitting in the audience, and as I watched his face I knew that would never forget that my power to kill was no less than my power to heal.

A good student of the history of medicine would know that the 18th and 19th century, in Europe primarily, scientific studies were taking place in the fields of bacteriology, histology, pathology, physiology, embryology and anatomy. Unfortunately, this was not the case in the United States, and except for a few major Universities, the medical schools had become diploma mills and had created a population of doctors, and a pharmacopoeia of poisons, that in 1860 provoked Oliver Wendell Holmes into saying ".....if the whole Materia Medica, as now used, could be sunk to the bottom of the sea, it would be all the better for mankind and all the worse for the fishes".

L.J. Henderson commented somewhere between 1910 and 1920 that "In this country, a random patient with a random disease, consulting a doctor chosen at random, had for the first time in the history of mankind, a better than 50-50 chance of profiting from the encounter". Although the physicians in Europe were better trained, the scientific community had yet to create the wisdom conducive to the rise of good therapeutic judgment, and the treatment related to it.

It was long in coming, but by 1930 the diploma mills had been closed in response to the Flexnor report of 1910, and well trained M.D.s were coming out of our professional schools. They brought into the small and large cities a type of wisdom never before seen in this country.

It was in the 1800s and very early 1900s, during a severe Flu epidemic, it was noted that those people who called themselves healers, and who were less invasive than physicians in treating this mortal disease, had better results and less deaths than the physicians of that time. It not only awakened the medical profession to its inadequate scientific base, but it encouraged Alternative care healers to grow in number and popularity. Thus it can be seen that Alternative cares did not grow out of a vacuum, during the 18th and 19th century, and now it is no different. 40% of the medical dollars in the year 2001 went to a growing number of Alternative care healers.

I was in practice only 5 years when I became aware of pharmacologic marketing and the techniques they used to seduce physicians into prescribing their new drugs. They suggested that the addition of a new molecule to the old drugs had improved their effectiveness, and had successfully counteracted the moderate toxic effects of the old drugs that were now being reported in the medical journals. It was not long before it became apparent that new drugs were pronounced to have minimal side effects, and the old drugs that had been out on the market long enough for the patent to be no longer protective, were discovered to have unacceptable side effects.

If you read the journals carefully you noted two overt facts. The articles written by outstanding men in the Universities had their research subsidized by the very companies that were producing the drugs. The companies therefore not only subsidized the physicians involved in the research, but they also subsidized the journals with heavy advertisement budgets. We physicians, who knew none of the researchers, but were highly dependent on their studies, were clueless as to whether all of these subsidies effected the outcome of the research. With evidence of the eventual and frequent changes in the recommended drugs, the lessons of Dr. Knapp and my dear old cardiac, clearly presented itself once again to my growing skepticism.

This was the evidence that made me realize that Managed Care had truly started when I was a young and gullible neophyte. However, it didn't drive me out of medicine until insurance companies, HMOs, lawyers, and non-medical secretaries were telling me how to treat my patients. They had the power to effect the destiny of those whose illnesses were fragile and potentially mortal. For many years I had moderate control over the new and old drugs I prescribed, by overseeing my therapy very carefully. This was not the case with the new HMO restrictions and I chose to leave medicine and see in what way I might affect the delivery system if I gave it deliberate and considered thought during my senior years.

Many critiques have been written on today's medical delivery system. In 1977 Dr. John Lantos wrote "Do We Still Need Doctors?" It was a brilliant, scathing discussion of the many problems in medicine today. Just like many of the critical books and essays on medicine that have been written in the last 100 years, there have been few suggestions regarding a cure for the ills they've described. Now that I've retired, and after 50 years of practice and nearly as much bitching, I am now ready to approach the possible resolution to many of our medical crises. But first, from this Lantos' book, I would like to list the crises as Dr. Lantos saw them, and add a bit of my own observations.

1) The problems of Informed consent- truth and honesty vs white and harmless lies.
2) The loss of touch with intimate medicine in favor of technical and Reductionist medicine. I have considered it the difference between Wholistic and Partistic medicine- the treatment of a human being or just a sick part.
3) The imposition of a third party payer- whether it be an insurance company, an HMO, a PPO, or the federal or state government.
4) The hierarchic distribution of medical care in this country, so that in terms of our success as a Nation in treating our entire population, we are 37th on the list of world countries and first on the list of costs, or a % of GNP that's over 16%.
5) The doctor is no longer the final decision maker as to treatment options and must relinquish control to a legally constituted, but non vulnerable, profit motivated corporation while the M.D. is still legally vulnerable.
6) Technical Medicine has an inflated value, while cognitive and touch medicine has been demoted. The direction in which a young student in training goes is somewhat determined by his medical school debt, and the type of life he chooses to lead. Suddenly money, rather than human compassion, becomes the underlying force that motivates the future of most graduating physician.

11

Wallace Salzman

These changes in medicine and the evils they represent are only a beginning of the list of causes for the negative changes that have taken place in medicine. Although the physician has pointed his critical fingers at lawyers, insurance companies, HMOs, politicians and a host of other professions eager to get a piece of the growing economic pie that the health professions represent, it is important to examine the physician himself to understand what role he might have played in bringing this parasitic infestation of the noble profession on itself. What happened to the role and image of the physician over the last 100 years?

The list of questions and answers are uncountable, and even with the completion of this book, questions grow at a rate that seems greater than we can solve them. I am very aware of the weaknesses of my attempt at organizing a means of solving these problems, but as Herman Wouk offered in his dedication to "This Is My God", he relied on the maxim of Rabbi Tarfon in the Ethics of the Fathers.

"The work is not yours to finish, but neither are you free to take no part in it."

Wallace Salzman

Part 1

Chapter 1

As you drive into Ortho-Para Park, on the far Eastern tip of Long Island, to your right you can see a beautiful green lawn decorated with islands of multi-color flowers. In the distance, the angry sea is crashing onto black dioritic boulders. The tidal action is so powerful that the under-tow has ravenously eaten up the sandy beach, and the battle between the sea and land is only quieted on rare occasion when the ocean suddenly turns into a tranquil lake. If quiet for a long enough period of time, the sand and the silt settle and a cobalt blue Long Island Sound reveals its serene beauty and beckons those who are fine swimmers to dive off of the rocky cliffs into the cool waters of the Atlantic.

To your left is a large parking area, with plantings located strategically throughout, and when you return your focus to the reason why you came here in the first place, you see, off in the distance, a piece of artistic and extravagant architecture known by most as the Medical Research Institute.

"The Ben Yosif Center for Ortho-Para Research" was established approximately eight years after the death of Dr.Ben Yosif. It was financed by Nathan Ben Yosif Jr., the grandson of Dr. Ben. Due to tragic circumstances, Nathan had never had the privilege of knowing and meeting his grandfather. This was caused by the violent antipathy between Dora Yosif, Dr. Ben's wife, and Ben at the time of their divorce. Her anger drove her out of her home town, and with their three children she never made contact with the doctor again.

During Ben's lifetime he was unable to trace their where-abouts, so that his children, Stanley, Zelda and Nathan grew to maturity totally unaware of their father's existence. It was only when Nathan Jr., Ben's grandson, grew to his seniority that he heard mention of his doctor grandfather by way of Helene

Lewis, Ben's nurse for fifty years, and who ultimately became his life long companion.

With instructions from Helene, Nathan eventually got in touch with Jonathan Fentonowsky, Ben's prize student, and had the privilege of hearing the story of his grandfather's life.

Nathan was a successful broker in Manhattan's Wall Street District, and had made a multi-million dollar killing in the market. When located by Helene, he was moderately depressed because the market no longer intrigued him. At 35 years of age he did not know what to do with the immense wealth he had gained in the 15 years he had mastered the Future's market. With his intuitive investments in Genetic stocks, just before they escalated in value, he made another killing when the cloning of organs was successfully accomplished.

The meeting of Jonathan and Nathan at Helene's home was at the time that the Ortho-Para test was finally accepted by the Medical Community. It was to be the beginning of a new life for Nathan who had lost his dad 15 years before, and was very much in need of a mature male on whom he could bestow his need for the father figure he had never had.

When Nathan approached Helene's home he was nervous at the prospects of meeting his grandfather's nurse and Jonathan Fentonowsky, who was considered Ben's Protégé. He got out of the car and stood back, surveying the home that had housed his grandfather. It was evening and the lights on the porch were lit brightly. He rang the bell.

It was only an anxious moment before Helene opened the door and looked with absolute joy at the attractive face she was seeing. "By God Nathan, you are the image of your grandfather when I met him for the first time nearly 50 years ago. Let me look at you. Now I know the true meaning of deja vue." She approached him with a gracious smile and before Nathan could open his mouth she embraced him as if he were her long lost child.

"I assume you're Mrs. Lewis" he muttered, when he was finally given a chance to speak.

"Yes I am Nathan, and you're a precious gift for my eyes and ears. I am seeing my darling Ben when I look at you. You are as he was when I became his nurse nearly 50 years ago. Oh, how I wish he had had an opportunity to see you so many years ago."

"And I him," Nathan replied. "You must tell me everything about him."

Jonathan was standing back during this emotional meeting between Helene and Nathan. He was surveying Ben's grandson from a distance and then approached him with a look of amazement on his face. "So you must be the Nathan that Helene has not stopped talking about since she talked to you on the phone. I'm Jonathan, the man who sat at the feet of your marvelous grandfather. He was without doubt the most brilliant and successful physician since Aristotle. I loved him and owe my premature success to his guidance from the day I entered Medical School until the day he died only 8 years ago.

Without hesitating, Jonathan went over to Nathan and lovingly embraced him as he would a long lost son. The emotional honesty of the greeting was foreign to Nathan who had never personally encountered this behavior before, and he automatically stiffened and withdrew enough for Jonathan to quickly release him and recognize that Nathan was not Ben, and their relationship could not be on a fast track.

Nathan immediately recognized what he had done and realized he was with a group of people who were different from his immediate family. There was an honest sincerity in the touch and body language of everyone in the room, and he had yet to meet the beautiful woman who was standing at a distance, gleaming what appeared to him a personal joy in seeing him.

Jonathan quickly stepped back. "Please excuse me for attacking you Nathan. I was a little out of control on seeing your uncanny resemblance to your grandfather. Please, I want you to meet my wife Helen who loved Ben possibly just an infinitesimal less than Helene, who spent nearly 50 years of her life catering to Ben's occasional cantankerous behavior."

For some strange reason, that he didn't understand, Nathan was beginning to feel an inner joy that was foreign to him. "Helen, I'm so happy to meet you all and I can see I'm going to learn more about grandpa Ben than I ever could have learned from grandma Dora. It seems as if, in their separation nearly 32 years ago, they both walked very separate paths and experienced their journey in very different ways." He smiled as he spoke.

Jonathan listened to his slow delivery, his careful choosing of words and was amazed not only at the amount of physical similarity to Ben, but also the carryover in speech and the cautious behavior he had demonstrated at their first meeting. He imagined that if he were asked a difficult question, he would close his eyes, as Ben did, and carefully mull it over before he would open them again and begin a long dissertation on the subject in contention.

Helene interrupted. "Here we are standing at the door. Come into the living room, relax and I'll get some wine to celebrate the beginning of a new Ben Yosif era."

They all followed her into a modest size living room, well aged and filled with old furniture and surrounded with shelves that were lined with hundreds of books and magazines that were the reflections of the searching mind of Nathan's grandpa. Helene had changed nothing after Ben's death and the prime evidence of her attachment to her mate was a large portrait of Ben taken when he was in his mid-fifty's.

Nathan was immediately drawn to this portrait. He studied it for quite a while, as everyone else quietly observed him. He then turned to face them, with tears in his eyes as he spoke, "Now I know why you were so shocked when I walked into this house. If ever there was evidence of the power of genetic structuring, it is the similarity of grandpa's and my body structure and the amazing postural similarities and facial emotions that I thought were purely learned. Now I can see genomic forces at work."

Jonathan laughed. "Nathan, I feel the same as you. The similarities are amazing, but your description of the similarities, are not the verbiage of a Wall Street professional. Have you had any biological training?"

Helen chimed in "I was going to ask the same question, but only in a different way. Where did you get your medical training?"

They were now seated and Helene was bringing out a decanter of white wine and was filling the glasses. Nathan was now comfortable with these strangers and knew there would be a lot of questions that had to be answered before the evening was over. He responded to the last question. "You are both right. I finished high school a little earlier than usual. I was only 16, and with an accelerated schedule at Columbia University in New York, I finished my pre-medical requirements when I was a little older than 18. I was on a fast track and entered Medical School very young and enjoyed the didactics of the first two years very much. I began to casually invest in the market in my third year and luckily hit a huge pay-load with each investment, so that by the time I finished my third clinical year I had a quarter of a million dollars working for me. I found the market far too fascinating, more so than Medicine, and I'm sure it was because I was only 20 years old. I foolishly left Medical School and went to Wall Street where my intuitive strengths continued to make me a fortune.

I am now 35. Money is no longer the attraction that it once was 15 years ago. I'm ready to use it for a worthy goal that can take me back to my initial track, one chosen when I went to Medical School."

Jonathan was intrigued. "That means that you'd only have one more year of Medical School and a residency to finish before your second goal had been successfully accomplished."

"That's right Jonathan and I've already made contact with the Cardwell School of Medicine. They told me that if I successfully completed my Board exams, which are a review of the didactic first two years, they'll matriculate me and assist me in finishing my last year. It was not difficult to convince them when I offered them 1 million dollars for their research in genetic engineering."

Everyone began laughing. "Nathan" Helene asked. "Did you give them the money contingent on their accepting you back?" She was gleaming in approval.

Nathan smiled, "Of course, Helene. I like to know the answers, before I make requests."

The mood was high and Helen turned to Jonathan. "Honey, Nathan would have fit very comfortably into your grade school class with the other 17, and probably would have made it even more interesting." She then turned to

Nathan and explained the experiment that Jonathan had created nearly thirty years previously. All of his class was now about the same age as Nathan and were now all prominent leaders in their respective fields.

Helene was the first to ask. "When do you begin your fourth year at Cardwell?"

"Actually I began last week. I'm taking my clinical year at a hospital attached to the Johnsonian Foundation in New York. Are you familiar with that organization Jonathan?"

The room erupted in laughter. Jonathan quickly answered. "Nathan, I was head of the Ortho-Para Psychology department for 20 years and only left two years ago to build a private Institute in Long Island. For that matter I just bought a 100 acre piece of property where I plan to build a Research Institute for our studies in Ortho-Para Medicine." There was a huge smile on Jonathan's face.

Nathan suddenly became serious. "I want to tell you a little bit about Ben's family, before I make you an offer."

The mood quickly changed. They all sensed that the gayety was to soon become serious. Nathan continued. "When my grandmother Dora left Ben, she went to the Blue Ridge Mountains in Virginia. She was flush with the settlement she got with her divorce and was going to live the life that Ben would not allow her to live. Dora was a child who wanted to play and Ben, I understand, was a gifted thinker who didn't know how to play.

She bought a beautiful home in the mountains, away from civilization and brought up her three children rather poorly, as is attested by her three subsequent marriages, the endless parties she was always yearning for, and the heavy drinking that ultimately became a regular part of her daily life.

My father, Nathan Sr., began drinking at an early age and had no guidance from the many men whom Dora chose as mates during her very sad life. She died from liver failure at about the same time that my dad died of lung cancer 15 years ago.

My aunt Zelda and uncle Stanley are still alive, but the Blue Ridge mountain experience contaminated all of their lives and I no longer keep in touch with them.

My dad Nathan married my mom Sally about two years before I was born, 37 years ago. I was their only child. When my dad died 15 years ago, she went into a one year depression, and thankfully one morning she awakened and was renewed. She decided to volunteer at our local hospital and has been head of the volunteers for the past 12 years. She never remarried for fear that she was unable to make a sensible choice after the alcoholic choice she originally made, but she has met some wonderful men who have become an important part of her life. Mom needn't worry about money with the trust that I've created for her, and she is now happily living alone. Her work, her friends and my success have been all she needed during these latter years of her life.

Helene, you'll notice that I didn't drink your wine. After watching my dad and his brother battle their drinking habit, at a very young age I decided that I might have the gene that helped destroy him, my grandmother and uncle Stanley. Please forgive me for being overly careful."

Helene interrupted him. "For heaven sake Nathan, you needn't apologize. You look beautiful. You've taken good care of yourself and your Mom. You really are a wonderful over-achiever, if I've ever seen one, and I guess I should have expected that with Ben's genes in you. I think you had a need to cathars yourself and I'm glad you did. I only hope that your sad beginnings will now come to an end as you follow in the footsteps of your wonderful grandfather."

Nathan graciously accepted these comments and turned to Helen. "How did you and Jonathan meet, Helen?"

With a broad grin she began the saga of her courtship with Jon. As she completed the story she added "Jon and I have discovered the formula for loving and living and how important both are to health and ultimate balance. Do you have anyone special in your life?"

Nathan appeared a little embarrassed by the question "Well, Helen, No. Not at present. I did on several occasions, but when I shared the joy I felt, when I was in the presence of these women, my Mom discovered their dangerous faults and I broke up with them as politely as I could."

Helen quickly understood the pathologic position his mother was in, but said nothing regarding it. "Nathan, you still have plenty of time and I know when you meet some of the yet unmarried women of our 17, you might thank Mom for making you wait." Jonathan thought Helen was brilliant in the way she answered that part of the conversation. "I agree with you Helen. What do you think about introducing him to Laura?"

"Jonathan shush. You were not designed to be a marriage broker. For that matter Nathan has proven himself to be quite a brilliant broker himself. I'm sure he will find a wonderful mate someday."

Nathan smiled but he was more prone to have a serious face. He hesitated for a few moments and then turned to Jonathan and stood up. "Jonathan, I want to make a deal with you. Please listen and don't stop me. On that 100 acre piece of land I want you to put up your research center in honor of my grandfather Ben Yosif. I want it to be named the "Ben Yosif Center for Ortho-Para Research." I want it to be completed within three years and if you don't have an architect, I know one who I feel knows as much about the problems of Medical Architecture as any man in this country. I'll introduce you to Daniel Namzlas, of Syntech Corporation, at the right time.

If you do this for me I'm in the position to give you, as a first donation, 30 million dollars. If that's not enough, I'm sure we can, without too much difficulty, find you more. I want the project to be completed within three years when I finish my training."

Jonathan, Helen and Helene just sat there in shock. Although this was part of their dream, it was a long way off and in truth it was to be a gift to their youngest son Ben Yosif Fentonowsky, who was now only ten years. It would be quite a while before little Ben graduated from his formal professional education.

When Nathan had finished, Helen was the first to speak. "This is a deja vue. It is like 25 years ago when the Worth Brothers donated millions to the University after my accident. Only this time I didn't need to break my hip to get it." She was a little hysterical with laughter as she was talking.

Jonathan remained serious. "Nathan, are you sure you want to be so generous after knowing us for only 2 hours?"

Nathan's answer was in rapid fire. "Yes. I've never been more sure as to where I'm putting my money. Understand, although I didn't know my grandfather, I've known of your Ortho-Para theory since it was first published in the Medical journals. I've known of your association with the Johnsonian Foundation and that is why I chose to go there. I know, without a second thought, that this will be the best investment I have or will ever make. So, please accept it without fear that I am being too quick in my decision making."

Nathan walked over to Jonathan and held out his hand. Jonathan stood up and to everyone's surprise Nathan embraced him in a way that let everyone know he was now part of the Ortho-Para family.

Chapter 2

It was not unusual for Dan, the chief architect of Syntech Corp, to spend nearly as much time with a functional analysis of a structure he was designing, as on the final structural drawings he would present to the Project Director, and the Institute Board of Directors who made the final decision.

Jonathan was chairman of the Institute Board and he personally had chosen Daniel Namzlas because of his extensive hospital experience. It was amazing that Nathan had made the same recommendation. The project that Dan was designing would be more than a hospital, since it would be incorporating not only the Research facilities for the Ortho-Para Foundation, but also facilitate a greater understanding of the relationship between Endocrine, Autonomic and Central Nervous System function, and in what way they controlled the multi-systemic functions of the meta-cellular community that made up Man.

Since Dan refused to design any structure, without understanding the functional necessities of every space, he created a giant problem for Jonathan who was unaware of the needs of the many Complimentary Medical Professionals he planned to have practice in this all encompassing facility.

The first meeting between Daniel and Jonathan was exciting for both of them. It had been three years since the 17 had gathered to celebrate the successful publishing of the fourth Ortho-Para Treatise that advanced the means by which the numerous Alternative Care procedures worked on some of the chronic degenerative diseases that Allopathic Medicine just palliated and failed to cure. Shortly after entering the Medical office, Daniel was ushered into Jonathan's suite where they held hands and embraced.

"Jonathan, my secretary told me we would be working together. What are we supposed to be doing? I haven't prepared any presentation, as yet." Dan had a questioning look on his face.

Jonathan quickly responded "I'm going to have you create the most unique Medical Facility in the United States and I needed someone who understood the laws of Contraction and Expansion, and Ortho and Para behavioral patterns, to do the job." As he was speaking he didn't project any evidence that he was joking, and Jonathan's seriousness alerted Dan that he was on the verge of creating his first major project. It would be the first attempt at going beyond the traditional, but he had yet to have described to him exactly how big the project was. "Dan, the project will incorporate space for all of the Alternative Care professionals to practice their trade."

"That's a large order Jon, and I don't quite know how I can structurally create the unique ambience you're requesting without a complete functional analysis of what would be done in each area we design."

Jonathan scratched his head before answering. "At this point I can't answer the functional questions you need answered before you start your creative work, but I wanted to have this preliminary discussion with you so that you're aware of the immensity of the project. I want to take you to lunch, since it's already 11:30, and after that we'll drive to Long Island where I'll show you the land on which our dream project will grow." Jonathan seemed quite calm, but it was not difficult to recognize an element of concern regarding the requirements for the structure he had in mind.

"Wonderful Jon, I'm ready for lunch and I'm curious to see where you're placing your dream institute." Daniel had now relaxed and was readying himself to tackle a project that would be unique, in that it would reflect on concepts that had never before been housed together in brick and mortar.

"Tell me Jon" he asked "Just how many Alternative Delivery systems do you propose to house in one place?"

Jon hesitated for a moment and then smiled. "In truth Dan, we are dealing with one system, the medical system. The question remains how many parts are there to this system, and which deserves a legitimate place in a Research Institution? Every semi-mystic has a notion, but it doesn't necessarily give birth to a legitimate specialty that warrants serious study. Many of the now existent multi-disciplined doctors claim to use Alternative approaches to their patients, but with the powerful Mental factors that can give nearly 62% placebo affects with many of the therapies, I believe we'll eventually be able to narrow down the functional needs of the Institute considerably, without losing any of the perceptual or inceptual therapies being offered." Jon's questioning face, as he answered, provoked Dan to wonder.

"What do you mean by perceptual and inceptual therapies Jon, and what truly is the placebo effect?"

During this conversation they were driving to a local sandwich shop and for a moment, when they had gotten their food, they stopped talking and settled down to eating. Dan began laughing, as he asked a question, at the same time he was watching Jon take a big bite out of a lox and cream cheese bagel delight.

"Jon, you haven't changed. I remember when I sat at your feet 26 years ago. You used to eat that same sandwich with the same look of delight on your face after every bite." They both chuckled as they reminisced about the days when Jon taught the class of 17 about the laws of Systems and Particles.

After a few minutes without talking, Jon responded to Dan's question. "A perceptual therapy is one that depends on whether one or more perceptions are part of the therapy. That would include the effect of visual, aural, tactile, olfactory, lingual and any other therapy applied from the peri-environmental space about the patient. In a sense it is the outside world physically touching on our perceptions and having a beneficial effect on our health.

An inceptual therapy is a response of the outside world to an inner need; a deprivation that is being fulfilled. Eating would be the prime example. When you are hungry you eat. When you are thirsty you drink. They are the essential examples of deprivation therapy. But there are more subtle hungers, such as needing a trusting friend; seeing your mate who is returning from a trip; a visit from your child who has been off to camp or college. These are all internal vacuums, which when not filled, create a "longing" that persists and at times has a devastating effect, if not handled properly by the caretaker."

Dan listened carefully and eventually took out a small pad and jotted down some notes. "As I understand you Jon, when the adversarial world of the outside is non adversarial and gentle, and when the inner world of need and deprivation is not being deprived, we have a summation of how many of the Alternative practitioners may achieve their therapeutic effects."

Jon laughed. "Yes Daniel. You've taken my verbal explanation and summed it up beautifully. But as you spoke, I realized we may not find the functional summary you need as difficult as we both initially thought. I'll send you a detailed list of the Alternative therapies and what I think they'll need in the way of unusual and specialized equipment. As far as the many subspecialties of Allopathic Medicine are concerned, I assume your hospital Architectural experience has taken care of that part of your designing problem.

"You're right Jon. That is my forte, although the speed at which the medical technical industries are inventing new products, it is sometimes overwhelming. In a hospital I did in California, by the time I finished the hospital, a new PET scanning technique had been developed. It was a teaching lesson. You must make room for tomorrow or by the time you finish the structure, its functional excellence has been already compromised."

Jon thought for a moment. "That's what troubles me Dan. Things happen. The creative chore I've given you will be breaking ground for a new integrated approach to Medicine. It has never been done before. You have no-one to talk to. The very fact that "Things Happen" is the danger in any pioneering project. You are it, and what you design will be the prototype for all other comparable Institutes that I'm sure will be popping up all over the country and eventually the world."

As Jon spoke he realized all the more how sensitive were the responsibilities he and Dan were assuming.

They were finished with their meal and were off on a one hour trip to the site that Jon had chosen for his dream Institute.

Chapter 3

Five years had past since Bertha, Jonathan's mom, had her first bout with the symptoms of brain dysfunction. Jonathan was visiting her in Baltimore from his home in New York. As he got off the plane, she was visibly choked up, revealing feelings she had previously hidden deep in her soul. It welled up to volcanic, eruptive proportions as she approached him and then, just as quickly as these feelings revealed themselves, they disappeared.

One small moment she dared to reveal in public that furnace of passion she choked on each day. This was not the Bertha that Jonathan knew most of his life. This sensitive loving lady, within a year's period of time, suddenly found the need to protect her vulnerability and to do so, she acquired a tantalum facial shell that frowned and smiled at will, but had lost honest spontaneity.

This was not the loving mother that Jonathan had gotten used to. In its place was a projective, angry lady who had become unhappy with her family and had developed a public and private face, the public emanations of pure charm, wit, and graciousness that endeared her to strangers, and the other, the private one, that was designed to attack her family who she claimed did not appreciate and love her the way they should. She refused to see a doctor to discover a reason for her sudden personality change. She saw no change and felt that Jonathan, Helen and the children were in need of medical help for ignoring her the way they did.

Helen was getting more and more upset as she observed the dynamics of her family, and she finally spoke up, "Jonathan, something is wrong. Mother has changed so radically. I'm afraid she has developed some mental disturbance, a senile dementia or possibly a brain tumor."

It was not long after Helen revealed her fears, while Bertha was walking down the stairs at home, she tripped and tumbled down a whole flight of stairs, striking her head on the landing. In the emergency room, at the Johns Hopkins

Hospital, an x-ray and CAT scan of her head revealed a large area of mid-brain necrosis, thus finally giving a reason for her marked change in personality.

Jonathan had her immediately hospitalized. After further studies her Neurologist was still unsure of the etiology of the necrosis, but thought it might possibly be due to one of the prion diseases Jonathan winced. " Are you telling me that the prognosis for a change in her anger and negativity is very poor.".

"Right. We can hope, but I'm doubtful. The necrotic changes have been growing for a long time."

Jonathan was beside himself. "She never complained about headaches. She just changed from an angel into an angry witch."

When Bertha left the hospital, she was sure the doctors were trying to kill her, and emphasized the fact that her doctor son did not try to protect her from those brutes.

Jonathan, with head bowed, shared with Helen his dismay. "Helen, I can't get angry at her anymore. She can't subdue that negative assault on me and those she loves. It is a selective Paranoia that we'll have to live with, unless we can convince her to take some medication, but I'm not sure what we can give her without over-sedating her and changing my angry mom into a depressed non-functional robot."

Helen was confronted with a problem she never anticipated. "What will we do Jonathan? We are both too busy to bring her home and nurse her. Although she is still up and about, we can't expect her to live alone. Jonathan, our whole future life has to be re-thought so that Mom is safe and possibly made comfortable for the months and years to come."

Jon quickly responded. "Helen, your negative thinking is premature. She was taking care of herself prior to her hospitalization and it's quite possible she'll not only be able to, but she'll insist on living alone. I know Mom and a hanging-on, dependent person she has never been, and she'd be uncomfortable unless she made her own decisions."

As Jonathan was speaking, he thought adoringly of Helen, who even before he had registered concern regarding his Mom, was already talking about the family's responsibility to take care of her. Nuclear family and its cohesive bonds and the traditions that bound it, had suffered a great deal in the ten years that preceded Bertha's illness. No less than the fissioned nucleus of the atom and cell, in our autonomy minded world, everyone seemed to be experimenting with loosely knit networks of egocentric design and of transient, self serving worth, while at the same time they seemed to be turning their backs on family and its allocentric traditions. Both Helen and Jonathan knew that the experiment would fail, but many people like Mom would suffer the loss of this traditional gift because they lived at the wrong time.

Helen responded. "I understand Jon. I just wanted to be sure that if she is unable to do the job herself, you have my complete agreement as to what our responsibility must be. We are a family and I've given a lot of thought regarding our family system. When we are no longer willing to commit our energies to the system, it will weaken and die. A system that is succored by the young and middle age has the power, in its very potential, to serve the needs of the very young and very old. It then has a sense of future, present and past and becomes the stable, relational bond that ultimately serves the needs of us all. It therefore services the functions that allow for both individual and family longevity."

They both slept fitfully that night.

Helen was awake at 5:30 and by the time she had showered and dressed, Jon was ready to enter the bathroom and groom himself for the day. They had gone to sleep very late, and although there was a sense of excitement regarding the new project and the wonderful young people who were part of it, Jon was also dealing with a serious personal problem.

By the time sleep was settling in, a new revelation suddenly awakened Jon. The Neurologist was right. This was a prion disease and in some cases it was inheritable. It was a small, proteinaceous, infective particle that caused a series of neurologic disturbances called Spongiform Encephalopathies. They attacked animals and humans, and dependent on what part of the brain was infected, the symptom complexes were given different names. In animals the most known of the encephalopathies was Mad Cow Disease. It suddenly became clear to Jon that Bertha had the Gerstmann- Straussler- Scheinker Syndrome. It had initially revealed its presence with evidence of marked imbalance, which in time proved to be a form of Cerebellar ataxia. Initially the etiology was not recognized, but the progression of the disease and the absence of any of the normal pathogens that caused similar symptoms, plus the MRI and Cat scans of the brain which showed marked necrosis, made Jonathan realize the diagnosis was pretty certain. Only a biopsy of the brain and a chemical analysis would ultimately prove the existence of the prion protein.

Bertha was 70 years old when the diagnosis was first made. The disease soon spread to the motor areas of the brain, and now in her fourth year of the disease, she had difficulty moving and comprehending the scope of her disability.

As Helen and Jonathan were sitting, eating breakfast, the phone rang. Helen answered the phone. "Hello----What's happening? ---- Jonathan and I will be there shortly.---Thank you." There was a look of frustration on her face. "What is it?" Jon asked.

"It's mom. This is the seventh day she has been unable to swallow and she's now non responsive. The nurses feel she needs I.V. fluids or a gastrotomy."

"To hell with that!" Jon shot back angrily. "She's dying and I'm not going to prolong the dying process with any invasive therapy. I'm sure she's suffering

and she'll need some sedation, if the disease itself has not obtunded her sufficiently to just put her to sleep until she passes away." His initial anger quickly disappeared and a full awareness of what was now happening cleared his
thinking and helped him make a final decision.

"Helen, it's very close to the end now. I'm going to call the staff at the office
and let them know I'll not be back until mom is gone. I'm not sure how long it
will take. I'm going to cancel my appointments right now and I'll leave it up to
you, as to how you'll handle your schedule.

Helen quickly responded. "First, I want to call George at college and let him
know what's happening. I'll then call the office and will join you with Ben at
the nursing home as soon as I can."

This sudden crisis was not unexpected. Bertha's deterioration had been progressive and she already had beaten the odds on survival, which was usually
two years and she was now in the fourth year of her illness. Little is known
about the prion diseases except the unalterable, non treatable, slow brain
destruction, ending always in death.

It was particularly difficult for Jonathan to be facing the treatment of an
infectious disease for which he could do absolutely nothing. His knowledge of
the entire concept of prions began in 1950 when a disease known as Kuru was
found to be a prion disease that occurred in epidemic proportions in the high-
lands of New Guinea. It was eventually found to be due to a soup, which the
natives made from ground up brains. Apparently a brain from someone who
had the disease was fed to many of the villagers and they all became infected.
Eventually when this soup was stopped, the epidemic ceased. At the time of
this prion epidemic, no therapy appeared to be effective and that is the case
today.

Jon reached the nursing home before Helen and found his mother in coma,
awakening only for moments. She would then fall back to sleep and had long
moments of apnea, followed by short moments of hyperventilation. It was the
breathing of incipient death. When he entered her room it was dark, with only
the sounds of labored breathing, and the odor of uriniferous decay. Life was
struggling its final battle with only hopes that the final dream, the hereafter that
every soul reaches out for, actually exists.

She was being watched by the whole family. When Helen arrived, little Ben
asked Jonathan "How is Grandma, Daddy?"

"She is very sick darling and she is dying. But she's not in pain and sleeps
most of the time."

Ben, with tears in his eyes responded "But Dad, you're a doctor. Don't let her
die. I don't want her to die. Why is she dying?"

"Ben, my son, it is because she is very old and her body is not working well
most of the time."

"Dad, what is dying like?"

"Dying is going to sleep and never feeling pain or sadness again. For Grandma it is not sad, because otherwise she would be suffering. It is only sad for us, because we love her and want to play with her again. But Ben, you've seen that each week we've visited her, she hasn't changed. She can't talk to us anymore. Isn't that right Ben?"

"Yes daddy. Will she be happy?"

"I think she'll be happier than she's been for a long time."

Ben was satisfied. He was now ready to lose his grandma, as long as she would no longer be suffering.

Bertha was therefore not alone. Her son Jonathan was sitting at the bedside singing his repertoire of songs in English, Hebrew, Yiddish, German and French, songs he had learned when he was a child. The only communication between mother and son was a hand squeeze that only took place if he stopped his singing. Bertha's eyes were now wide open and fixed. She stared frighteningly at the world of the living, holding on and not yet ready to join her husband Sam who had died many years before.

"Twinkle, twinkle little star,

How I wonder what you are.

Up above the world so high

Like a diamond in the sky."

"Mom do you remember teaching me this song?" Jonathan teasingly asked his mom, but there was just a miniscule squeeze of her hand, as she held onto his tenaciously, never letting go.

"Row, row, row your boat gently down the stream,

Merrily, merrily, merrily, merrily life is but a dream."

All of the 50+ years of Jon's exposure to music was an open catharsis that would not end until he had successfully eased her through the door from her present hell on earth, to the peace of eternal sleep. Helen sat by Bertha constantly rearranging the covers and pillows, as evidence of her restlessness and discomfort revealed itself.

"Sh'ma Yisroel, Adonai Elohenu, Adonai Echod."

"Hear O Israel, the Lord our God, the Lord is One."

Bertha's breathing became more stertorous and there were longer periods when her breathing stopped, and then suddenly began again with a quickening of her ventilation.

"Frere Jacqua, Frere Jacqua

Dormez vous, dormez vous."

Two hours had now passed and with every song there was a squeeze of her hand. It was Jonathan's assurance that she heard and wanted to continue to hear. Only two weeks before, Jonathan had walked into the Nursing home and was greeted joyfully by a demented Bertha.

"Oh Jonathan, you're such a lucky boy. You are now my prince. I was just crowned Queen of West Virginia. Do you realize Jonathan, I am a Queen and you are a Prince."

All of the nurses were happily reinforcing this happy delusion, and the angry lines of Bertha's face were temporarily transformed into pure joy. This was the last moment of pure joy that Bertha would experience, as her dying heart was slowly beating muffled marches to the grave.

Jonathan was now watching her last battle, as he helplessly sat by witnessing the battle between the Prince of Darkness and Bertha Fentonowsky, who showed no signs of being ready to sever her ties with her darling son.

All of these past 5 years of memories went through Jonathan and Helen's minds, as they watched Bertha in extremis, now beginning to battle for each breath as she weakened from her long battle, and eventually from the superimposed symptoms of starvation and dehydration that began only days after she was crowned Queen. Jonathan revisited his decision not to feed her and thus allow his precious mother to die the way she had lived, battling for the right to do it her way.

As he watched his mother, Jonathan imagined his dad watching from his celestial seat, cheering her on in her last mortal battle with the forces of death. "Come on Bertha. Give it your all. Let the world know how you lived each day of your life from the moment you awoke."

Even as Jon vicariously sensed her dying pain, he was proud of this fighter whom he called Mom for the 55 years that had passed since his Baltimore birth to the young and beautiful Bertha and Sam.

The fleeting thoughts of Helen were no different, as she found herself reliving the death watch she experienced first with her Mom, and later, her Dad, just before she entered college. She remembered a letter she received from her former principle, Mr. Ross, when she lost her dad.

Dear Helen,
Jonathan just called and shared your pain with me. My wife and I extend our sympathies to you and your family. Courageous George is now busily clearing the path for all of us and I'm sure our own celestial journey will be easier because he went first. Peace has now displaced the horrible technical and painful invasion of his firm body and mind, and after hearing what he went through in the last few hours of life, I bless his death for having detached him from the busy, frantic and hard working doctors who unfortunately look at death as their enemy instead, when the patient is suffering, as an ally and friend.

Only God knows whether death can truly be more painful than life, especial - ly when we live with grief, with losses, and with fears regarding the future, as George did when your mother died.

But grieve we must. The tears relieve the spirit and also let George know, as

he momentarily looks back at what he left, that he was loved and was signifi -
cant to you all and that he'll be missed by those who loved him the most.
  Give my love to Jonathan and I hope to see you both soon.

Mr. Ross

Whereas Sam Fentonowsky and George Prentice were accepting and understanding of their wives occasional irrational outbursts, both women grew wiser and remained unfettered by convention, ethnic or cultural, so that they were exciting spirits to watch as they invented and designed new ways of approaching old problems. Helen realized that the Judaic origins of both mothers had produced offspring, she and Jonathan, with genes that were perfect for the introduction of the new and argumentative theories that had become the base of Jon's and her life.

Bertha was now the last to go, and she was fighting a furious battle.

"Mom, squeeze my hand" Jonathan pleaded, as her breathing became more stertorous and her color ashen blue. For the first time there was no response. She finally seemed to be detached and no longer fighting to remain attached to Jonathan's and Helen's world.

Through tears, Jonathan pleaded with Bertha, "Mom, it's alright. You can close your eyes and let go. Dad is waiting for you."

With a wide pursing of her lips, she suddenly responded to his plea, stopped breathing, closed her eyes, and passed through the Golden Arch of Heaven.

Jonathan and Helen lay their heads on Bertha's body and began crying quietly. They did not know that their sons, George and Ben, were standing at the door watching their parents say good-bye to grandma.

Bertha's death was at sunset on Friday. It was the beginning of the Jewish Sabbath. Services would have to wait till Monday and would be followed by her cremation. Sam's ashes had been part of Bertha's household for 15 years, and she haunted Jonathan, when she knew her death was imminent, to be sure her ashes were mixed with Sam's and then distributed back to the Earth as Jonathan and Helen deemed fit.

It was Ben's first experience with death. At 10 years of age he was already intellectually far ahead of Jonathan's 4th grade class of 16 years before, but he was emotionally following the genetic schedule associated with a much slower Embryonal evolvement process. Whereas he enjoyed the games and playmates of his own age, he was already reading at the level of college sophomores, to the concern of his parents who knew that soon they would be unable to assist him in many of the subjects that fascinated him. For the remainder of his life he would have to seek his intellectual goals with minimal help from his parents.

George, on the other hand, was already 20. His intellectual evolvement was not precocious, but his involvement with music and the arts had reached near spiritual heights, as his piano skills took on the challenges of the great masters. His hands had elevated the piano repertoire to heights only achieved by some of the great impresarios of the day. It was scary. To Jonathan and Helen's surprise they had, in their two sons, representatives of the two extremes of Ortho and Para, and although they felt proud of their accomplishments and the accolades they received from their friends and the public, they were very aware of the dangers that such polar extremes represented as they got older.

"Mom," Ben asked "with all of my readings, I never came across a description of human cremation. Is there a special machine that does it? Can we watch it happen? I know that the Vikings put dead heroes on a boat or raft, set it on fire and sent them out to sea."

Helen responded "No Ben. Grandma will be placed in a crematorium, a large furnace. All that will be left of her will be some gray and white ashes, no different than those of Grandpa Sam's that you examined once when you visited Grandma Bertha.

"It really makes a lot of sense" Ben responded, "taking the ashes of grandma and grandpa and spreading them in a garden of the home that you and dad will be spending the major part of your lives. Are we living in that home right now?"

"I'd think we'd have to ask dad, but I doubt it. It is not impossible that we'll be moving many times, when you and your brother move away from home and are involved with your own careers. We won't need as large a home and we'll probably look for something smaller."

Ben then smiled as he presented what he thought was the answer to the problem. "Then Mom, we have to go to a National Park, like the Grand Tetons, and toss their ashes into one of the lakes or gardens so that George and I can visit them every few years, and our children and grand-children will always know where they can touch on their heritage and at the same time enjoy the magnificence of Nature in a great Park."

Helen was used to hearing wonderful answers to difficult problems from her Ben, and took it in stride. Although there were brief moments that Jonathan sought the quiet of his study to ramble through past memories and grieve the loss of his Mom, he was too aware of the past few years of suffering that had thankfully come to an end. Just as his Mom and Dad had relished watching their son struggle with the problems of life, it was now Jon and Helen's turn to stand by their sons, who were already wrestling with the resistances to the achievement of excellence in their respective fields. It was in the outside world where Passion and Reason met their greatest personal obstacles.

It was now the Sabbath and Jonathan and his family were sitting their first day of Shiva. It would last all of the seven days, when Jonathan and his family would remain at home and receive condolences from the crowd of friends who had become part of their lives during their many years of residence. There was a natural limitation to this reception. Bertha lived in Baltimore and the Shiva would be held in Bertha's home, while the majority of Helen's and Jonathan's friends were in Ohio, near the University, and in New York, near the Johnsonian Foundation. Helen felt it would limit the number of visitors; would make the crowd easier to handle, and she knew that Jonathan had already emotionally adjusted to his mom's death because she had suffered for so many years. She and Jonathan were sure that only their Baltimore friends would be present at the funeral service.

Both Saturday and Sunday passed quickly, with visits from Bertha and Sam's friends, but the numbers had already dwindled over the years because of their age. They were mostly frail women in their 80's, who were less spry than the Bubbly Bertha who remained the catalyst for excitement up to the moment her brain began to fail her.

It gave Jonathan plenty of time to write a Eulogy to his mom, so that her formal send off would be worthy of this exciting lady who played so powerful a role in preparing her son for the role he was supposed to play one day, in the field of Medicine.

Chapter 4

Orpstains Funeral Home was conveniently located only two miles from Bertha's home. Helen awakened early on Monday and happily witnessed the sun rising in the east. The cloudless sky foretold good tidings for the day. Jonathan would bid his final farewell to the seed carrying roots of the Fentonowsky family. Sam had no brothers or sisters and happily bestowed the privilege of preserving the family heritage on Jonathan when he bid his final farewells so many years ago. Jonathan considered this a privileged responsibility that he had faithfully fulfilled with the birth of George and then Ben.

Helen walked about the house quietly so as not to awaken Jon. He had stayed up most of the night writing his mom's Eulogy and was still sleeping soundly. Most of the direct neighbors had already visited and offered their condolences, so that he expected the audience at the funeral home to not be large, as it would probably be filled only with the locals.

He awakened at 7:00 and shuffled into the bathroom, still not clear of mind, but ready to take on the difficult day.

"Helen" he shouted, "good morning dear. What do you think I should wear?"

"I don't think you have a big choice Jon. The black suit with the faint pin stripe should be suitable. It's been recently cleaned and you look very nice in it." She smiled as she answered, because she remembered a remark Jon made when he took off the suit the last time he wore it and complained about his growing pouch.

"You'd think that I had had the two children instead of you. Look at my abdomen growing. Helen, I think it's time for me to get a little more physical. How about tonight baby."

She answered "Exercise, but not on me you horny brute." But in recent weeks, during the final days of Bertha's illness, intimacy never entered Jonathan's mind.

"Jonathan, did you finish your writing last night?"

"Yes, I think I'm done, but I'll have to re-read it after breakfast. What time do we have to be at the funeral home?"

Helen thought just a moment before she answered. "We'll be receiving guests beginning at 10:00 A.M. and the services will begin at 11:00. I told everyone who called, there would not be a funeral procession to any cemetery, so that they could plan the latter part of the day without concerns regarding those services.

Jonathan was pleased. "I'm really tired honey and I don't regret not going to a cemetery service. I initially worried about cremation, but the rabbi registered no objection. I'm sure Mom had checked on that before she put it in her Will."

At 9:30 they left for the funeral home and were shocked to see the parking lot already full. Jonathan was at first surprised and turned to Helen.

"They must have several funerals going concomitantly. I'm glad there is a parking space for us."

They checked Ben to see if he was properly groomed. They saw George's car was already parked and they felt ready to make their final farewell to Bertha.

As they began walking toward the entrance, they saw a familiar face driving into the parking lot. Jonathan was the first to see him. "My God Helen, that's Mr. Ross driving in. He must now be the same age as Mom."

Helen had a big smile on her face and didn't look a bit surprised. When Jonathan saw her smile he quickly guessed that Helen had something to do with his being present.

"You called Mr. Ross, didn't you?"

Helen's eyebrows raised, "Yes I did and I called a lot of other people who I knew would be angry at us if we had not. Let's go inside before you register anger at my uncalled for behavior."

That quickly brought an end to any critical remarks from Jon and they entered the Chapel that had more than 100 familiar faces ready to greet him. It brought immediate tears to Jonathan's eyes. Helen turned to him and Spoke, "Bertha has been part of every celebration we've ever had. Every one of these people knew Bertha as a close friend, not only as the queen who was your Mom. I didn't beg these people to honor Mom. I merely told them she passed away and they insisted on knowing the time and place of the funeral."

Jonathan quickly got control and went about greeting the class of 17, all of whom were now powerful professionals in their respective fields. His friends from the University and the Johnsonian Foundation had come to say their final good-bye to Bertha, who at one time had hugged and kissed every one of them at one or more of the parties celebrating the success of her son and daughter in law.

The moment of the final services arrived very quickly and the Rabbi took his place in front of the closed coffin and waited until all was still.

34

"I have already blessed the path to glory for dear Bertha who has wished for death to my ears only, since she lost her Sam. It is unfortunate that her dying process was so filled with suffering, but there was a terrible ambivalence that created this tragedy. Sam was drawing her to him, and Bertha did not want to leave Jonathan and his family. Thus we had the push and pull between wanting to die and wanting to live, which touches, to a degree, on all of us sitting at the bier of this wonderful mother and grandmother. Although her dementia increased over these past few years, when she saw me and knew I would officiate at her funeral, she seemed to brighten up for a few minutes and made an unusual request of me, and although Jonathan will give today's Eulogy, the things I speak of now, could only be said when she was gone. These words were to be expressed to Jonathan and his family. I will read them.

Dear Children,
I am so sorry that I was so uncontrolled, and that I couldn't stop the terri -
ble, angry words that came out of my mouth every time you came to visit me.
Strange as it may seem, I felt better when I catharsed my anger. There was no-
one else I knew who would understand this need other than my Jonathan and
Helen. Even in my final days you gave me the freedom to fight my demons and
relieve my body of the pent up anxiety that wanted to make me even crazier
than I appeared most of the time to you. When the Rabbi reads this I will be at
peace, holding the hands of my darling Sam. We will both be happily smiling
as we look out over this sad service, knowing that we will cause you no more
pain. My darlings- Love each other and remember me to all our dear friends-
Bertha."

The audience was silent. It was an unusual addition to a funeral service, but what else could you expect from Bertha?

The Rabbi continued. "I'm sure the good Lord will find dear Bertha a marvelous addition to his special minyan, when the more exciting services are held on the celestial sabbath. She was blessed on Earth and I'm sure she will be blessed in heaven."

The Rabbi then turned to Jonathan. "I believe it's your turn to bless your Mom."

Jonathan rose and approached the pulpit. He was wearing a shawl and touched its fringes on the casket, kissed it and then faced his friends.

The Eulogy

Who else would think of sending a letter of apology to be read after her death? My Mom was the magic cross between the incorporative, nurturing woman of the past, who saw her biggest role in life was to have children, to nurture them until they had created a sound base from which they could take off and ultimately do anything they chose to do. Their choices would be carefully genderized, so that both male and female propensities and competences would always be carefully taken into consideration during the training process.

Recognizing the importance of the role of motherhood, and having given it comparable marks to the most complex roles that man or woman can now play in our modern society, Bertha strongly believed in the gender differential, and in her family life she was fortunate enough to have Sam as her husband, who stood by her and never challenged her female powers, no more than Bertha challenged his male prerogatives. She firmly believed that the job of running a home, of nurturing and training a family, was comparable in difficulty to running a Nation, and if women did a better job in raising their families, we would have more men and women capable of performing their specialty roles in their chosen fields of Medicine, Law and Philosophy, and in the more general roles of parent, teacher, politician and preacher.

All my dad had to do was recognize Mom's authority in the home and he needn't worry about his authority outside the home. It was a fair division of responsibilities and because my Dad never demeaned my Mom, she grew into the very complete nurturer that she became.

Who would know better than her son, who watched her suffer the agonies associated with the mothering of a hyperactive male child?

This was my Mom.

1) She nourished me by way of her uterus and umbilical cord. She lived a non toxic life during her pregnancy with me.

2) She then transferred the nurturing responsibilities to her breasts, her warm loving arms, and the rest of her body which was in constant motion in service to her son and her mate. That was my Mom. She was exemplary of what a woman should be.

3) For the first six years of my life mom taught me, socialized me and praised every movement that I made, so that my powerful self esteem was attached to the relationship between Mom and myself. In a nutshell, she was ready to assume the responsibility of being the architect of her child's unfolding while preparing me for my first out of home challenges when I entered school.

4) For Bertha, school was not an end point to her maternal responsibilities, and until I reached the age of twelve when I entered Junior High school, she was in constant touch with my teachers and carefully oversaw the decisions I made, as I assiduously tried to break free from her nurturing, which to my young mind was comparable to a chain preventing me from some wild experimentation that I was not ready for. As my experiments with life appeared to her more dangerous, her voice grew louder and she began to use the services of my Dad who had a more authoritative social position in my life. Needless to say my freedoms grew directly in relation to my competence in handling such freedom, and it was not I who made

the judgments regarding my competence. It was Bertha. It was she who let me know when I was ready to go to college and would no longer be under her command. She knew that if she did a good job, after seventeen years of surveillance that I could be anything I wanted to be.

My darling mother was right. I became just what I wanted to be, and as she watched me fulfill my dreams she would tell my Dad, occasionally within my hearing. "See what a good job Jonathan is doing. Kiss me and say thank you."

When exposed to college and away from Mom, I was contaminated by the social antipathies that filled the campus and I eventually changed my name to John Fenton, to the dismay of my darling parents. That lasted exactly two years, and for the rest of their lives my huge guilt was a source of my energy as I tried to neutralize that idiotic moment of my life, and seek their forgiveness.

This Eulogy is my confession. It is the confession that relates to my adult childhood, when I thought I was ready to take on the challenges that only a balanced adult could do correctly.

What mark do I give you Mom? You were terrific. You were an A+ woman who understood your difficult role, as you toiled with your overactive son.

What mark should womankind give you?

An A+, for having discovered that women and men have very different roles in life, and it is dependent on the society in which you must act out the labors of your life that you become what you choose to be.

Your soul will be greeted by the righteous committee that attends to the gates of heaven, and I'm sure they will show you where dad has been patiently waiting for you. Give him our best regards and we'll be hopefully coming your way in the distant future. I love you."

The Chapel remained still until Jonathan left the pulpit, and just as prior to the service, a line formed to meet Jonathan, Helen and their children.

Those people who had visited Bertha's home during the first three days of sitting Shiva, quickly left, so that only those who had traveled from distant cities remained to talk to them. There was a third group of young adults who remained seated and did not proceed to the waiting line.

Jonathan personally thanked the many professionals who had traveled from the University, in Ohio, and New York, to honor his mom. They were all sensitive to this period of grieving, so that they spent only a few moments embracing their friend and within one hour Jonathan turned to face his former students of 26 years, who were now 35 years old and already prominent in their respective professions, anxious to touch base with their friend, their mentor and the scientist who had helped them choose the paths to extraordinarily successful careers.

Jonathan approached them, eager to touch, embrace and hear from them all. He automatically turned to David. "Do you think that everyone can come to

Mom's home where we can spend the remainder of the day? Does anyone have plane connections before 5 o'clock"?

No one responded.

"Then come to my home. We'll feed you and you can feed me my Manna, which is simply a review of all of your lives since I last saw you."

Helen was standing next to him and gleaming her delight at seeing all of them. "You are all that we have, to partially negate the sadness of this day. In truth, Bertha was ready to die and Jonathan and I, after seeing her suffer for too many years, were ready to let her go.

Life can sure present itself in marked contrasts. You are just beginning the exciting, knowledgeable portion of your careers, as Bertha completed hers. We should pray that we do as good a job with our goals, and responsibilities, as she did. Come on. Let's go to Bertha's home. We could pay no greater tribute to that great lady than to discuss the future of this great country, from each of your respective professions."

It was an afternoon of sadness and joy for Jonathan and Helen. When one loses a suffering parent, there is an element of relief that momentarily displaces the grief. But like all the other vacillations of life, the shifting emotions only abate with the passage of time. The presence of his students made at least the moment more bearable for Jonathan, and therefore more tolerable for Helen and the children.

When they had all partaken in a delicatessen lunch, Jonathan's formal searching began.

"Daniel, I want you to share with your colleagues the most recent project that has been offered to you." As he spoke he had a huge smile on his face as the challenge was presented to Daniel.

Daniel's face quickly opened up and without hesitation he turned to his colleagues and began to outline the project that had been financed by Nathan Ben Yosif. He went into detail regarding the unfortunate circumstances that ultimately molded Nathan into a brilliant financier and now a new doctor in the making.

"I would assume that the 30 million dollars that Nathan advanced towards the creation of the Ortho-Para Research center is only an eighth of the ultimate cost of this immense project.

Are any of you familiar with any of the various Alternative Health Care products that are available to the public? It is claimed that 40% of health care dollars are now going to the many Alternative care systems now available."

Laura was the first to respond. "You know that I've been working with Helen on the Ortho-Para base that under-girds all of the health care systems. The number of patients we've seen who are involved with different systems of care is enormous, so that the evidence that the patient population in this country is becoming somewhat disenchanted with western Medicine, at least as it's practiced in this country, is enormous. Your statement that 40% of health care dollars are now going to the Alternative cares, now sometimes called Complementary care or CAM, is amazing because so much comes right out of the pockets of the patient who is not covered by insurance.

It is just this frightening trend that concerns me and I wondered in what way could the physician and the medical delivery system, be responsible for this trend?"

Allison quickly responded. "All you have to do, to answer that question, is to survey the history of Medicine in this country. As I read the details, I realized that the authors of this history had their own prejudices that were somewhat hidden in their statistics. As I recall, in the early 1900's there were about 130 Medical Schools, whose specified entrance requirements were non existent. The teaching staff was only practicing physicians made up of the graduates from these particular schools. There were inadequate books available for their studies so that following a written report in about 1910 by Abraham Flexner, a great number of the Medical Schools shut down.

There was far from unanimous agreement amongst physicians. Flexner rightly complained about the absence of the basic sciences from the curriculum, a fact which was not present in European Medical Schools, where the best doctors in this country were trained. I couldn't believe that basic science was not taught, but when you think of how young modern chemistry and physiology are, compared to the age of Medicine, it is sad but understandable. Thus the Flexner report maintained that science and laboratory medicine was essential to the training of doctors and recommended that didacticians in the various sciences, rather than clinical physicians, should be teaching at least the first two years of Medicine.

The objections came from men like Sir William Osler and Harvey Cushing who represented the best in American Medicine at that time. I remember reading Osler's objections at that time, so I brought a copy of it to read to you.

'Cabined, cribbed, confined within the four walls of a hospital, practicing the fugitive and cloistered virtues of a clinical monk, how shall he forsooth, train men of a race, the dust of heat of which he knows nothing, and this is a possibility, cares less? I cannot imagine anything more subversive to the highest ideal of a clinical school then to hand over young men, who are to be our best practitioners, to a group of teachers who are ex officio out of touch with the conditions under which these young men will live.'

Osler was a man who participated in the dramas of his patient's lives and was able to reflect this very well in his opinion of those who would dare to teach that which they were not living. However, in the name of progress, and, Oh, how this concept has been misused in history, the relentless process of technical evolution took place and the salaried didactic scientist took over that very essential clinical two years of our students education.

Fortunately, that was not the final decision, because during the training of Kenneth, Cassy and I, Clinical Medicine, taught by practicing physicians, was added back to our curriculum during the first two years.

It was in this way that Technical Medicine was born, and Touch Medicine was witnessing the birth of a force that would ultimately contend for dominance. I believe that Laura was correct in suggesting that the Flexner report was probably the formal introduction of Science to the medical school curriculum. However, Laura, in Europe, they were far ahead of us and one of the reasons why the criticisms of American Medicine reached a loud crescendo, relates to some of the giants in Medicine who were willing to speak their minds. Oliver Wendell Holmes stated his firm belief 'that with the exception of opium and anesthetics, and a very few other drugs, if the whole Materia Medica, as then used, could be sunk to the bottom of the sea, it would be all the better for mankind and all the worse for the fish.' That was back in 1860.

What is more significant is that although Sir Osler was the embodiment of Touch Medicine, his textbook "The Principles and Practice of Medicine" contained very few treatments that we consider useful today, and many that were potentially lethal. Among the armamentarium of therapy were blood letting, the application of leeches, blistering, precordial ice bags, and the insertion of tubes in the legs to reduce edema. This was a primary treatment for cardiac diseases. True, they didn't have effective diuretics for heart failure, but when they did they began to poison cardiacs with mercury."

Jonathan decided to enter the debate and direct the attention elsewhere. "What you're saying is that the Medicine of the early 20th century was not the same as it is today. But please remember that only recently the New England Journal of Medicine reported 100,000 hospital deaths associated with drugs that were presumably given properly, with proper indications, and it was idiosyncratic reactions that led to death. It is just these facts that are making the frightened public re-examine the Alternative Medical techniques that are less invasive and therefore potentially less toxic."

Cassy questioned "Is that why you've decided to add the Complimentary therapies to your Research Center, Jonathan?"

Jonathan was slow in replying. "That is one of the reasons Cassy. First off, I want you all to recognize that Ortho-Para therapy is very much a part of the Complimentary therapy we're talking about, so that the Ortho-Para Research Institute that Daniel is designing, will not stand alone as an approach to medical illnesses, but will hopefully give meaning and understanding to so many different approaches to illness that have had healing powers which were unachievable by the traditional Allopathic physicians."

Daniel was next to speak, "Now all of you know why, when we knew we were meeting at Jonathan's Mom's funeral, I asked you to prepare for a discussion on Alternative Medicine. I understand the function and structure of general and specialty hospitals, but I'm unfamiliar with the functional needs of the many therapies that Jonathan wants to incorporate in his Research Center. Everything all of you can offer that will enlighten me on any of these healing arts will make this project more likely to be successful."

Becky was quietly listening to the discussion, but felt uneasy regarding the direction it was going and finally responded to Daniel's request.

"Dan, I realize that you expect little from the clergy regarding Alternative Medicine, but I'm going to surprise you. First off we've been under the influence of the Ortho-Para concept for the past 26 years, so that our thoughts and perceptions are easily focused on anything that might resemble the concept that Jonathan proposed so long ago.

In my Congregation I have a lovely Chinese couple with 5 children. He practices Chinese Traditional Medicine and during my early anxieties regarding the problems I had, when I began my pastorate, I decided to see Si Shu. He's a man in his sixties and has a degree in Western Medicine from Harvard, and a PhD in Chinese Medicine from the Beijing University. His practice is overflowing and from the first moment I sat in consultation with him, I was introduced to Yin Yang."

Becky turned to all of her classmates with an air of astonishment. "I swear to you that this 4000 year old concept, that was finally made a part of Chinese Traditional Medicine in 200 B.C.E., is so similar to the Ortho-Para concept, it shocked me. The Autonomic Nervous System was 2000 years from being discovered and understood, and yet on a basis of the observations of human behavior, biologic, earthly conformations and experiences, they created the very under-girding of a Medicine that at times mirrors Ortho and Para."

Her shock was shared by some of the class who had difficulty conceiving of such an intuitive miracle, and the look on their faces reflected their doubts. It was therefore appropriate for Jonathan to step in and monitor the discussion.

"I know it is difficult to believe, but absolutely everything Becky has told you is true. It was nearly 30 years after the Ortho-Para concept was first conceived by me that I heard of the terms Yin and Yang and I began my search into the history of Chinese Medicine. But it is now 4 o'clock and most of you have planes to make, so that it's best not to begin a critical analysis of the differences between 4000 year old Yin and Yang, and the 26+ year old concept of Ortho and Para. I must write a brief conceptual summary of Chinese Traditional Medicine for Dan, and I'll send you all a copy when it's complete.

I'd now like to thank you all for coming to Mom's departure celebration. It was really a celebration, considering the suffering her path of death took. I've been watching all of your achievements. I search the literature and have enjoyed watching your names coming up frequently in the fields that you've begun to influence. I notice Justin that you made a big switch since we last spoke. I believe it was at Ben Yosif's funeral."

Justin smiled brightly as he responded to Jonathan. "Yes Jon. After I talked to you, I decided neither of the two options I presented to you would be the direction I would choose, and I went back to the University and got a PhD in Bioengineering. I now work for a Medical Engineering Corp. that designs Bio-medical instruments for new approaches to surgical intervention. It's been very fulfilling, but I've yet to come up with anything that is revolutionary. When I do, you'll be the first to hear of it."

It was obvious that Justin had made the right decision and the class began to mingle in preparation for their departure. Dan had rented a small school bus to facilitate the trip to the airport. It was a repeat of the many times Mr. Ross had done the same thing. It was about 45 minutes from Bertha's home and they needed at least a 30 minute lead time.

Kisses and embraces soon followed and they were all in the bus waving good-by to Jonathan and Helen who were by themselves for the first time that day.

Chapter 5

It was nearly three weeks later that Jonathan and Dan landed at O'Hare airport on the way to visiting Becky in the far western suburbs of Chicago. She was the pastor of a large Protestant Cathedral, after successfully ministering to a smaller church for nearly four years.

Becky was aware that Jonathan and Dan were visiting her so that she might introduce them to Si Shu, her physician, and the man she had spoken of during her visit to Bertha's home. She was thrilled that she was able to participate in the early stages of Dan's Architectural venture.

When she opened the door to greet Jon and Dan, Dan approached and embraced her, suggesting to Jon there was an intimacy existent between Becky and Dan that he had been unaware of. Dan had never suggested he was seriously involved with anyone, so that his greeting of Becky was a surprise. Jonathan quietly watched them, as he stood back waiting for their embrace to end. He knew that Dan was Jewish and as quickly as this thought rose to mind, he remembered that Helen and Kenneth were both protestant, and they both married Jewish mates. It was not the world of two generations ago. The concept of ecumenicalism was touching on marriages throughout the country. For that matter the higher the level of education reached by University students, the less they considered religious affiliations as an obstacle to potential happiness, and they focused more on love, compatibility, and the real life issues that they would face more successfully and completely if they worked together.

Becky radiated her joy in seeing both of them. "I'm so glad you made this trip and I'm overjoyed that I mentioned Si Shu when I was at your home. Jonathan, I know you'll be thrilled with his knowledge and you'll leave requiring no other consultant when it comes to Chinese Medicine. Dan will have all the information he needs for that part of his project."

Dan quietly responded. "I assume that Dr. Si Shu is aware that we'll need at least a day, or possibly even more of his time."

Becky smiled and answered quickly. "He is prepared to give you as much time as you need. He is thrilled with your project goals and suggested he might enjoy the opportunity of practicing in such an environment and consolidating his education with so varied a group of professionals."

Jonathan looked a bit surprised. "Did he actually share that possibility with you Becky?"

"Oh yes. He's been practicing now for nearly 31 years and if there is more to learn about human behavior and the workings of the body, he felt that your new concepts, examined in the environment of your Research Institute, would be the place for him to do a little more growing." She was now relaxed and glowing the joy she felt in their presence.

Jon and Dan sat down and enjoyed the tea that Becky brought as they relaxed from their plane flight. Jonathan spoke. "Dan we may get more out of this first venture than we planned. There are only a few physicians who've studied the various Oriental and Asian branches of Medicine on their own, but there are even fewer who are certified in both Western and Eastern Medicine and who've been out in practice for many years. Becky, you may have added a huge addition to the success of our Institute."

She smiled when she responded. "Jon, I'll bet, before you're through with this project, that everyone of the 17 students whom you taught and influenced these many years, will ultimately add a dimension of great worth to yours and Ben Yosif's gift to Medicine." She went over to Jon and kissed him on the cheek.

Jonathan was looking at a beautiful and confident Becky and he grabbed her hand and squeezed it. "If what you say becomes a reality, it will be the ultimate in happiness I can ever expect. I must admit that I would not have expected a minister of God to be the first significant benefactor to the creation of the ends we were seeking. By the way, let me assuage my curiosity. Have you two been seeing each other, unrelated to our class goals?" His look of curiosity was seen in the tightening of the tissues about his eyes and the growing smile on the outer margins of his mouth.

Dan began chuckling. "Jonathan, you saw something we really weren't trying to hide. Yes, Becky and I have become a thing and to a degree we have been distantly involved since we met again at Ben Yosif's funeral. We decided that we would not marry until we could comfortably settle down in one place and continue to participate in our career growth. You know Jonathan, you and Helen set a precedent with the amount of time you waited before you married, but we recognize at our age it can't be too much longer. We're now 35 and since we want children, it won't be more than 6 more months before we set the date."

Jonathan got up and grabbed Dan and then Becky with big hugs and they all settled down to bathe themselves in the special joy that was felt at that moment. After a bit of silence Jonathan spoke. "Becky, when did you schedule our first meeting with Si Shu?"

She responded without hesitation. "Dr. Si Shu took a one week vacation beginning tomorrow. We have an appointment at his home at 9:00 A.M. Until tomorrow I want you to make yourselves comfortable. We have several guest rooms in the vestry so that you'll both have a good night's sleep in a house of God. Only good things can happen after that."

At 9:00 A.M., the next morning, they rang the doorbell of a beautiful home owned by Dr. Si Shu. The door was answered by a man who appeared much younger than his 60 years of age. He was thin, with graying black hair, traditionally yellow skin, with a white, well trimmed goatee adorning a well chiseled chin. His skin appeared unwrinkled, so that in summary, he was a handsome, bright looking, radiant personality.

Jonathan could quickly see why Becky was so impressed for he felt the same feelings as she and the good doctor hadn't opened his mouth yet.

His smile ushered in his Oxfordian English greeting. "Welcome to my home. Ever since Rev. Rebecca told me of your theories, reputation and present goals, I've been anxious to meet you."

He held out his hand to both Jonathan and Daniel and graciously invited them into the living room which was a true cross between old China and Modern America. Although the pictures and paintings on the walls were of the European masters, the pottery, lamps, vases and trinkets were all from ancient China and were the memorabilia of many of his trips back to Beijing, to revisit the land of his ancestors.

Jonathan responded to his greeting, "Thank you Dr. Shu for allowing us to visit you, to sit at your feet, and become students of the ancient profession that Western Medicine has chosen to disregard."

Shu radiated peace, a trait that was part of his very therapeutic personality. Jonathan noted that this quality was a powerful part of the personality of every successful healer he had met in the past.

Shu spoke "You exaggerate my position Dr. Jonathan. It is I who will be the student. I'd like to begin my lesson by your telling me what happened at the very moment you became aware of your concept of Ortho-Para and its physiologic control of the behavior and personality of all of us?" Jon smiled "Well Dr. Shu, you sure don't waste any time. It's been a long time since that wonderful Eureka birthed itself in the mind of a young ambitious doctor. Note that I speak as a distant observer of myself, because so many of my discoveries were made by that fellow I used to be. But, be that as it may, I remember clearly the answer to your question. I was reading a book entitled 'Structural Psychology' by D. and K. Stanley-Jones. He was describing the appearance of an infant and its irritability just before feeding and then its pink, relaxed, tranquil state, after taking to the breast and finally falling asleep with a smile of eupeptic pleasure. The discomfort of Ortho dominance and the peace of Para dominance, flashed through my mind and it has never left in the last 20+ years."

Jonathan watched Shu's face and he could see his eyes light up with apparent enthusiasm as he became witness to Jon's sudden triumphant moment of discovery.

Jon reacted to the turn about in questioning. "Dr, Shu, you're turning the student-teacher role around and if you've read the material that Becky gave you on Ortho-Para theory, you're way ahead of me on the sharing of information."

Shu laughed and apologized. "You're right Dr. Jonathan and I apologize, but I'm fascinated over how you had a revelation of something that has existed in Chinese Medicine for over 4000 years." He bowed his head as he was talking.

"I can understand your amazement " Jon replied "but what I think is far more amazing is that without the knowledge of the physiologic functional anatomy of the hypothalamus, the ancient sages could have differentiated the powerful differences between the forces of day and night. It is these historic insights of your ancient elders that I want to learn. I want to also see in what way Chinese Medicine can be incorporated into an Integrated Medical System that is not purely technical, logical and reasonable, as is Western Medicine. Man is just not purely left brained in the way he acts out his life in the many systems in which he functions. He more frequently is tactile, intuitive, passionate and even irrational, as he acts out the dramas of life at home and in the intimacy of privacy. Those are right brain functions and are part of the powerful healer's therapeutic package that he offers to the ill."

Jonathan felt his thought processes would soon be challenged by this powerful oriental physician who radiated gentleness and acceptance, as he listened to every word that Jon expressed. Shu appeared very happy with Jon's presence and it slowly relaxed Jonathan who was more prone to be on his guard and always ready to protect himself. "Si Shu, shall I call you Shu or Si ? I want you to be comfortable,"

Shu responded "I think you'll be more comfortable with Shu, and may I call you Jon, and Daniel I'd like to refer to as Dan."

Dan answered "I think that Jon and Dan would be comfortable for all of us."

"That's fine," Shu replied "It should make our visits more comfortable, less rigid and I hope our get-togethers will always reflect the proper Yin and Yang of an intimate relationship. I think I should tell you a little bit about myself, so that you'll understand the genetic and cultural forces that eventually turned me into what I now am. But before I do that, I need your permission to add one more person to this study group."

Jonathan appeared surprised. "And who is that?"

Shu quickly responded "My son Shang. He is a practicing architect who has recently become interested in hospital architectural design. He is 32 and very much a beginner, but his attachment to Chinese Cultural traditions are a little less than I think they should be. The very things we'll be talking about could act as a cultural catalyst, and it is one of the reasons I responded so quickly to Rev. Rebecca's request."

Jonathan graciously responded. "Then, by all means, ask him to join us before we begin our research into Eastern and Western Medicine, their differences and their many similarities. We'll have representatives of four decades of cerebration focusing on the cultural diversity that we all represent."

Shu smiled and left the room. He had his son waiting in another part of the home and returned very quickly with Shang. In contrast to his father, he was tall and had more Caucasian features than Jonathan expected. Except for the definitive oriental slant to his eyes, his color and general demeanor was more Western than Eastern, and both Jon and Dan had to restrain their surprise. Shu however perceived their reaction and quickly clarified their surprise.

"Jon, Dan, I want you to meet my half-breed son Shang. He has the eyes of his father and the graciousness and configuration of his beautiful mother, who doesn't have an ounce of Chinese blood in her. From the moment Shang was born I saw the interesting genetic process that designed him into the magnificent young man that he has become."

Jonathan quickly understood Shu's paternal talents and knew that Shang would be a wonderful addition to their important conference. Shang quickly responded to his dad's introduction.

"Please excuse my father for giving so complete a description of my racial, cultural and genetic origins. It pleases him and it doesn't offend me. I think that it quickly clarifies the complexities of my bi-racial origins which I'm very proud of." Shang had a beautiful smile, and there was no question that he was loved, since his proud father appeared to beam at his presence.

Jonathan was very comfortable with Shang, and also with Shu who was not yet ready to begin what would be a mini conference on Eastern and Western Complimentary Medicine.

Dr. Shu immediately began the discussion. "Jon and Dan, I think it's important to first discuss the remarkable similarities between Ortho-Para and Yang-Yin. On the surface Yang and Ortho, and Yin and Para appear to be synonyms, and I thought that would be a good place to begin our understanding of the processes each represent to our respective cultures. How would you summarize their differences Jon?"

Jon willingly bowed to the older Shu who had taken over an authority role in their meeting. "I think the difference can best be summarized in the processes that are dominant in the night-day rhythms, as they control the breaking down of structure during the day and building up of structure during the night. Para is the autonomic polarity of reconstruction and of the building up of Potential Energy, whose structural configuration is templated by the Kinetic

Energy or the activities performed during the preceding day. Ortho is the autonomic polarity of the breaking down of structure (P.E.) into Energy (K.E.), to
add movement, strength, endurance, thought, problem solving and motivation
to the living processes during the day in the outside world.

In short, Ortho literally changes structure into process and Para rebuilds the
structure, when we contract into ourselves and live within an inner world
where consciousness and thought no longer exist, as we understand it."

Shu listened carefully and tried to relate this Americanized concept of Yin
and Yang to the traditions he had been taught. The correspondences were
remarkably similar, even though the ancient thought processes that conceived
of Yin-Yang were unattached to the yet to be discovered Autonomic Nervous
System. He responded to Jon. "To clarify this more completely, do you have a
column of properties that differentiate Ortho and Para that is comparable to
this one related to Yin-Yang?" Shu handed him a list of Yin-Yang correspondences.

YIN	YANG
Dark	Light
Earth	Heaven
Moon	Sun
Night	Day
Autumn	Spring
Winter	Summer
Cold	Hot
Female	Male
Slow	Fast
Down	Up
Inside	Outside
Water	Fire
Metal	Wood

A broad smile broke out on Jon's face. "If I show you a list of the differences
between Ortho and Para you would probably note that it is a modern extension
of Eastern antiquity. With our modern scientific awareness of the Cosmos, Life
and Matter, we can not only extrapolate the Ortho-Para dualism beyond Yin-
Yang, but also intrapolate, (if you'll excuse a new word), so that the intrusiveness of Ortho and the incorporativeness of Para can be shown to be present
even in the elements, molecules and mega-molecules that preceded life. Today
they make up the biologic anabolic and catabolic particles which are controlled

by the day and night rhythms of every system, which function as part of the whole we call Human. Let me show you a list of Ortho-Para differentials that can be equated to your Yin-Yang correspondence.

Ortho	Para
Short	Tall
Fast	Slow
Responds fast	Responds slowly
Highly volatile	Calm and serene
Distrustful	Trusting
Large System	Small systems
Creative	Artistic
Revolutionary	Clings to the known
Easily Frightened	Integrated and self sustaining
P.E. > K.E	K.E. > P.E.
cAMP	cGMP
Expansion	Contraction
Oxidation	Reduction
pH down	pH up
H+ up	H+ down
Acidosis	Alkalosis
Electropositive	Electronegative
Daytime	Night time
Serotonin	Melatonin
Break down	Build up
Repolarization	Depolarization
Intrusive	Incorporative
Partistic	Wholistic
Entropic	Non-entropic
Rational	Passionate
Responsive to outside World	Responsive to inside World
Thyroid	H.Growth Hormone
Catecholamines	Histamines
Adrenergic	Cholinergic
Beta agonists	Alpha agonist
Selfless	Selfish
Hostile	Loving
Telos	Eros

Ortho	Para
Social law	Natural law
Yang	Yin
Somatic	Visceral
Catabolism	Anabolism
Exothermic	Endothermic
Fission	Fusion
Evolvement	Involvement
Projection	Intrajection
Creative	Being Created
Adrenalin	Insulin
ACTH	HGH
Corticotropin	Somatomedins
ADH	

While Jon was watching the response in Shu's face he noted "I want you to know that, in truth, this is a very partial list. Living represents the homeostasis between Ortho and Para. We can actively live each day and recreate ourselves during the night. The living process is truly a body breakdown process."

Shu responded "It may seem remarkable to us now Jon, but even Confucius, who lived in the sixth century B.C., picked up on the significance of Yin-Yang and expanded on it, so that the philosophies of Confucianism, Taoism and Buddhism became the three major faiths that designed the ethical and moral codes that, to a great degree, are recognized today. Needless to say, just as Yin-Yang influenced Philosophy, Religion and Politics, it also had a primary influence on Medicine. What I don't understand is why it failed to influence Modern Western Medicine. Aristotle, in his writings, spoke of the major elements that influence all life and they are so similar to the thinking of the Chinese ancient Philosophers. He spoke of the four major elements- Air, Fire, Water and Earth. At the same time the Eastern thinkers were speaking of the Earth, Metal, Water, Wood and Fire. Although communication between the East and West was very limited, the Chinese 'Yin' and the Aristotelian 'Body', and the Chinese 'Yang' and the Aristotelian 'Mind' were structurally and functionally identical to 'Para' and 'Ortho'. It was the Church that took the concept of a shifting balance between Body and Mind and separated them, so that the masses believed that whereas the Body was subject to death and decay, the Mind, which was equated to the soul, would have an eternal life in heaven. Jon, what I'm trying to say is that the concept of balanced opposition, so firmly believed in by the Chinese, and the Yin-Yang homeostasis that was essential to good health, has always been present in both Eastern and Western thought but it failed to be properly verbalized in any of the Textbooks of Medicine, at least in these United States. I know that Cannon wrote extensively on homeostasis within the physiologic processes in the body, but the Medical Schools failed to emphasize the properties that you so wisely incorporated in your Ortho-Para list."

Jon listened carefully to Shu, while Dan and Shang quietly sat and listened to their mentors. "I never realized that the Body and Mind, as conceived by Aristotle, was truly the same as structure and function, and therefore no different than Yin-Yang and Ortho-Para polarization."

Shu responded "So how do you explain the loss of homeostasis in modern technical Medicine? It is Yang without Yin; Ortho without Para; Reason without Passion. Before all of the technical and invasive instrumentation had evolved, the Western Physician had no more and no less power than the Eastern Physician. He was ready and willing to engage himself emotionally in the problems of his patients. Via his concern, patience and ever presence, he developed the trust of the patients and could alter the fears of the ill, and turn a concern regarding death to a powerful goal for life. Only a trusting relationship could accomplish this shift in mental polarity from negativity to positivity."

Shu was obviously very prepared for this meeting with Jon and was expounding on Allopathic Medicine's biggest weakness. He thus far had touched on the very thoughts that provoked Jon's most recent studies and the reason why he was happy to meet Dr. Shu who had apparently been most aggressive in his search for a more complete Medicine than his training at Harvard had given him.

Jon appeared more relaxed as he responded to Shu's comments.

"Dr. Shu, I'm so pleased to hear your reference to Aristotle. I must admit that in the past I didn't relate 'Body and Mind' to Para and Ortho, but it is certainly a relevant interpretation. In reading Chinese Medicine, I see a strict attempt at organizing human structure and function, although somewhat differently than Western Medicine. I understand that no less than human structure dictates function, you must create an organized Medical structure in order to organize the mind of the physician so that he can practice Medicine. It is the structure that always dictates function, and in return, the cybernetic feedback influences the restructuring process. After the thousands of years that the process of cyber feedback altered the functional failures of an ancient structure, the Herbs, Acupuncture and Moxibustion, are now the primary tools of Eastern Medicine and were firmly entrenched until the late 1940's when antibiotics came into existence."

Shu listened quietly and allowed Jon to finish before he contested his apparent summary of Eastern Medicine. "Jon, I have trouble accepting your rather brief summary of what works in Eastern Medicine. You've left out too many important essential practices that are daily regimes in the armamentarium of every practitioner. In truth, the successes and failures in Eastern and Western Medicine were comparable until antibiotics and the more sophisticated surgical procedures came into existence. Needless to say, ignorance has played a big role in the history of both Eastern and Western Medicine, and it is because of this there are so many Alternative Practitioners successfully practicing in America, China and in every country in the world. You must understand Jon that the majority of the world is taken care of by Shamans, Witch Doctors, Hybrids of Modern Medicine and those Psycho-mystic practitioners who seem to successfully meet the needs of the populations they care for. One of the big questions we should be answering is how and why they so powerfully influence their cultural illnesses."

Shang suddenly responded to the problem being discussed. "I would like to add one thing Dad. There is one aspect of the problem that intimately relates to the Yin-Yang hypotheses that I learned in my basic course in brain physiology. There are approximately 30 studies on right and left brain function and how they relate to autonomic function. It is an example of how different cultures use language to express what they consider the apparent differentials of function of both sides of the brain. Knowing that I'd be sitting in on this conversation, I thought it relevant to bring the expression of these differences to our meeting.

	Right Brain	Left Brain
Bugabo	Bad soul (left handed)	Good soul (right handed)
Hindu	Integral Thought	Rational Thought
Freud	Primary Thought	Secondary Thought
Fenichel	Pictorial	Verbal and Linguistic
Chinese	Yin	Yang
Spearman	Spatial	Verbal
Hobbes	Unordered Thinking	Purposeful Thinking
Pavlov	First Signal System	Second Signal, language
C.S.Smith	Gross	Atomistic
Price	Synthetic or Concrete	Analytic or reductionist
Wilder	Geometric	Numerical
Head	Perceptual and non verbal	Symbolic or systematic
Goldstein	Concrete	Abstract
Reusch	Analogic and eidetic	Digital or discursive
Bateson and Jackson	Analogic	Digital
J.Z. Young	Map-Like	Abstract
W. James	Existential	Digital
Luria	Simultaneous	Successive

Levi-Strauss	Mythic	Positive
Bruner	Metaphoric	Rational
Akhilananda	Manas (Mind)	Buddhi (Intellect)
McCleave	Emotional	Rational
Gaub	The Heart	The mind

Shang continued. "This work was done by Arthur Deikman regarding Bimodal Consciousness, the dual polarity of action in relation to the outside world which he called the consciousness of explication and that of the receptive modality of consciousness, a tacit and sensuous mode that closes off the outside world and focuses only on one's within-ness, no different than the infant."

Jonathan suddenly recalled the articles and interjected "Shang, I read the series of articles you're talking about, and if I recall correctly he equated the polarities he was speaking of as due to the Orthosympathetic dominance essential to the state of striving, the achieving of personal goals and the obtaining of social rewards, as well as the avoidance of failure and pain. In Ortho dominance the perceptions become more sharply defined as does the epicritic definition of the outside world. Thus a variety of physiologic and psychologic processes develop at the same time as the development of a multi-dimensional unity that is required if one is to manipulate the environment.

In contrast, the receptive mode is a state organized about one's perceptions, which is our intake from the environment and the emotional affects that they precipitate. The sensory perceptual system is the dominant agency. The polarity is Parasympathetic dominant. Whereas the EEG of the Ortho mode tends toward Beta waves, the Para mode tends towards Alpha waves and in contrast to Ortho, in which the muscle tension is increased, in Para, muscle tension is diminished."

Dan now had the courage to interject his own interpretation of what was being said. "We might be approaching an answer to that which we came here to evaluate. What is fascinating is the reciprocity of modes. Increased muscle tension enhances Ortho or Yang perceptions. Decreased muscle tension enhances Para perceptions, so that the other consequences of the polarity changes can follow. It explains some of the effectiveness of Body-Touch therapy and it appears evident that one's very posture can influence modal dominance."

Jon was pleased that both Shang and Dan had entered the conversation. It was time to suggest the true cause of the success of so many of the Alternative Therapies and to go one step farther in the understanding of the marked differences between all Men and Women.

Chapter 6

Jonathan was pleased with the discussion thus far. He felt it was now time to shock Shu and the younger men with the revelation of a Eureka that he had not even shared with Helen. He first wanted to incubate it longer and be sure there were no flaws in his thinking. In truth, this Eureka would comfortably underlie the explanation for all of the various applications of Alternative Medicine and explain the powers of the physician who existed before the age of antibiotics and the technical devices that are now in prominence. He therefore was anxious to pursue this aspect of their conversation.

"Dr. Shu, I want to interfere with this wonderful conversation that seems to have pretty well established that there does exist a Neurologic component to both the Rational and Passionate expressions of a balanced antagonism that explains all the processes of life. What I want to concentrate on is the passionate aspects of life.

Spirituality and God are derived from the pre-rational feelings of the global child who senses at birth a yet undifferentiated river of feelings we now recognize as Infantile Anxiety. It is during this preverbal chaos that an unknown, omniscient, omnipotent figure arises, one who assuages our needs, protects and embraces us, loves and explores our bodies and securely and tranquilly wins over our hungers and thirsts. She has the power and time to help differentiate these feelings so that we soon learn the meaning of the feelings we initially did not understand. It is at this time, during this preverbal period of life, that this image of an all powerful, loving figure, who is real and not a spiritual fantasy, comes into existence.

This awareness of a God figure is created before we can speak or reason. It is an unspeakable awareness that seems to indicate that we will always be taken care of by this powerful representation of the outside world. We expect she will always be available to calm our fears, obtund our pain, embrace and subdue our anxieties, service our survival needs and tolerate our foolish and dangerous experiments, knowing always that she was the one who oversaw our creation and now would always be there to successfully counteract our destruction.

The circuitry of this cerebral awareness took place when feelings were primitive, very powerful, but not yet articulateable. It was the time that our passions were templating the Right Brain, the Subcortical Hypothalamus, the Hippocampus and the Amygdala, the very areas that contain all of the feelings that wreak havoc when we are faced with dangers. These are the areas that affect our vitality and excitement, when we've fulfilled our goals and responsibilities. When all of our needs have been met, love envelopes us and we can't help but cherish the bearer of Touch and Embrace, the Fulfiller of all our needs.

All of this happens and grows during the first two years of life. It designs the very same qualities that we have adorned our God with. Who, I ask you, helped create this passionate and non-verbal miracle?

Why Mom of course. To the infant God is Mom.

Spirituality and God are derived from these pre-rational feelings of the global child who senses that the pains of hunger, thirst, stool and urine control that come from within, have always required the outside world to resolve these intrinsic problems. So it is with one's mortality. It is a fear that comes from within and is best resolved by an outside force that can simply explain it away and introduce eternal life as a possibility."

Shu, Shang and Dan sat open-mouthed at this interpretation of the eternal fantasy that pursues all Men when faced with incipient death. God is the Fountainhead of all Religions and is the eternal, yet terminal prayer that we all express as the curtain descends on life. Does this explanation of the origin of the internal awareness of an outer power negate the existence of a Heavenly power that created the Cosmos and has the power over all life and death? It was a question that would require a deep search by all of them. Jon however, was not finished with his revelations.

"Before I hear your response to what I've shared with you, I'd like to go one step farther. It appears that sexuality is a perceptual searching process and eventually a rediscovering of the reverse of the birth process. The cataclysmic, oscillating, vibratory penile-vaginal storm that terminates in an orgasm, ushers in a moment of Para tranquility. We are in a warm, moist, uterine primal sea again. In the intimate sexual journey, the perceptual journey of a fetus being expelled is acted out in reverse. The orgasm represents the processes of uterine and vaginal contraction and the forceful extension of the fetus's body just before delivery.

Yet although incomprehensible, in the short journey from the uterus to the outside world, the infant has touched, tasted, smelled, and has been massaged by the rugated passageway through which it passed. The dark cavern, through which it passed, had already left powerful, non-verbal messages of comparison between the idealic world of peace and the total dependency from whence we all came, to the new world where we're given the responsibility for breathing and for leaving behind peace for chaos, darkness for light and a warm watery environment for a cool, dry atmosphere. You've been placed in the hands of unknown creatures who will be teaching you all you'll ever know for the next 12 years of life.

At no time will your brain lay claim to the virginity it knew before your birth. According to John Locke, human beings are born as clean slates, a tabula rasa, with our minds being 'empty tablets capable of receiving inner and outer imprints but having none prior to birth'. The truth of this statement can be argued on a basis of what we've shared with you.

For the next two years, and no time more striking than the first, the infant will be learning and filing non-verbal data regarding its internal feeling world. It will be much later, when its eyes mature, that it will introduce the perceptual outer world, introduce language, data and the meaning and significance of all the intrinsic feelings that caused the infantile chaos in the first place.

Their meaning will not only be made clear, but also the importance of their teacher, their Mom, who taught them by taking away the negative feelings of hunger, thirst, and wet, stinky diapers, and in their place made them comfortable with her embrace, warm breasts filled with milk and wonderful smells, thus eliminating the first feelings of chaos, and changing them into the recognizable, exciting world of adventure and discovery. Although different, it will be nearly as perfect as the tranquil sea from whence they came.

Each day designs the next day's competences. Mother will be responding to their every need and want and will continually talk to them, so that these needs are eventually given meaning and power and a means of one day communicating intelligently with this overwhelming but responsive outside world that is willing and ready to serve.

The only language with which we're born are our postural changes, our facial emotings and the verbal cries and grunts that add meaning to early language, and at times become part of the communication process for the remainder of our lives. Acceptance and rejection have an early means of communicating our feelings in this way.

Despite this preverbal awakening, those early internal and external experiences of life are never approachable via Psychologic and Psychiatric means, even if our life was pathologically lived, because the routes of language have not yet developed and the experiences that may have precipitated a wide array of unhealthy feelings, were not verbally attached to cause. If you examine tables 12 and 13 in the first Treatise, you will see the anlage of the Para and Ortho feelings and in the way they sequentially unfold during the inceptual (feeling) phase of early development. The concept of the pleasure/pain unfolding of an infant is far more complicated than the bipolar duality expressed by Freud.

These feelings, having been completely differentiated, are waiting for the time the perceptual apparatus is ready to introduce the outside world and give articulateable meaning to the infant. As the "Person, Place, or Thing" of outside world becomes a part of the child's perceptual awakening, the feeling tones, previously described, are attached to every object so that from the second year on, there is a subject-object fusion taking place, with every object having an emotional positive or negative cathexis. This underlies many of the problems associated with the Embryonal unfolding process, which requires, in the world of science and logic, a separation of the subject-object fusion. It is historically not until the Renaissance that this separation began to take place even in the average adult.

It is for these reasons that the linguistic, mental and rational sciences have failed, too frequently, to solve the mental illnesses that unfortunately grow in number each year and appear to be approachable only by pharmacologically interfering with the neurochemistry of the brain. Although not spoken of frequently, especially in relation to the treatment of psychotic illnesses, the drugs are used mainly to modify the patient's illnesses so that the caretakers can more easily care for these very sick people.

If these and other illnesses are not rationally approachable via the theories of Allopathic Medicine today, what can we do to discover a means of treating those problems in which the early developmental brain has gone astray and becomes the behavioral cause of problems early or later in life?

With this in mind I approached the problems, which seemed to be hiding behind closed doors and that in some way had to be opened. Just as the virgin brain of the infant pounded on the uterine door that would open the world of the outside and eventually teach language and the art of reason, it was my goal to pound my contaminated rational head on the pre-verbal, pre-rational doors of the already closed uterus, and contract into my post birth global brain where pre-verbal mysteries lie and discover the passageway back to the beginnings of life where the create/destroy ratio is high enough and illnesses of the mind and body do not exist.

Dr. Shu, the pounding process has taken nearly 26 years and the total by-product of this search, although far from complete, has altered my thoughts regarding human behavior, man's most innermost feelings, and its effect on the Allopathic Medicine I've been practicing for these 26 years."

The men sat still after Jon was apparently finished with this portion of his lecture. They realized, for the first time, Science, Reason and Intellect were not to be the only approach to Illness, especially since so many of the most severe problems that Medicine has failed to solve, were out of the Realm of Reason.

Shu was the first to respond to Jon. "I'm sure that your feelings, when this

Eureka surfaced, were no different than the original Greek Eureka when Archimedes discovered the laws of floatation and water displacement. I hope that you didn't run through the streets nude, although after a Eureka of that magnitude, I would have forgiven you for anything. Let's be sure I understand what you're saying. It is the birth process that awakens the perceptions to their first awareness of the frightening outside world, but it is an inceptual (or feeling) awakening first. It is your contention that the sexual journey is just a recapitulation of this same sequential process of birth, only in the opposite direction. Therefore I see you; then I hear your voice; I then get closer and touch you. If it meets with my approval and yours, I smell you and then I taste you. By that time, the touch is getting more intimate and it is at that time that genital contact begins the finalizing of the perceptual-inceptual journey. It ends only when the cataclysmic orgasm initiates the rebirth of the Para or Yin polarity from which we eventually came. It is for that reason we are born in Para and maintain a hold on the Para Veil. The visual and auditory introduction of Ortho does not truly take place until this veil is slowly dropped and we become more aware of the outside world. Unfortunately, we suddenly become aware that we are truly alone and no longer the master of our global world."

Jon smiled and went over to Shu and embraced him. "Wow. I think that your description was more precise than mine. Yes, and does it give you an awareness of what it means regarding Healing therapies?"

Shang was quick to respond. "You must excuse me for interfering in your dialogue, but this is too exciting for me to be still. I would like to answer you Dr. Jonathan. Every Alternative Therapy that I've heard of involves the stimulus of one of these perceptions and from what you've both described, we start from the distant visual spectrum and move slowly toward the ultimate genital contact which initiates the conjugal reflexes that ultimately create the highest state of Para polarity that we can obtain. Is that right Dad?"

Shu quickly agreed with his son. "If that is true, it is our responsibility as healers, above and beyond the writing of prescriptions, to get as intimate with our patients as we possibly can, without violating the marital or family system which we are treating. Shang, I asked you to bring a list of all the Alternative Therapies now being offered the public. Do you have it with you?"

"Yes Dad. Let me give a copy to all of you."

With that being said, Shang handed the following copy of the list to all of them, and remained silent while they read it.

The Alternative Therapies

Acupressure
Acupuncture
Aerobic exercises
Aikido therapy
Alexander techniques
Allopathic Medicine and
    its specialties
Aromatherapy
Art therapy
Aston patterning
Ayurvedic Medicine
Bioenergetics
Biofeedback therapy
Biomedicine
Body manipulation
Bone setting
Cell therapy
Chanting therapy
Charismatics
Chelation therapy
Chinese Medicine
Chiropracty
Cold Laser therapy
Color therapy
Colon therapy
Curanderismo
Dance therapy
Detoxification
Dietary Medicine
Divine Healing
Drumming
Electromagnetic therapy
Environmental Medicine
Enzyme therapy
Exercise therapy
Feldenkrais
Food therapy
Flower therapy
Guided Imagery
Heat therapy
Hellerwork

Specialty Medical therapy
Physical therapy
TNS
Psychiatric therapy
Radiation therapy

Chemotherapy
Immunotherapy
ENT therapy
Opthalmologic therapy
Pulmonary therapy
Cardio therapy
Genito-Urinary therapy
Gynecologic therapy
Transplant therapy
Bone marrow therapy
Hormone replacement
Antibiotic therapy
Gene therapy
Galvanic therapy
Electrolyte replacement
Allergic therapy
Orthopedic surgery
Neurologic surgery
Dermatology
Auditory therapy
Cranio-sacral therapy
Abdominal surgery
Rheumatology
Proctology
GI therapy
Infectious Disease
Parasitology
X-ray
CAT scans
MRI
Positron Emission T.
Chemistry Deter.

The Alternative Therapies (cont.)

Herbal therapy
Holistic Medicine
Homeopathy
Humor therapy
Hydrotherapy
Hyperthermic therapy
Hypnosis and Imagery
Hypothermic therapy
Juice therapy
Light therapy
Love therapy
Kinesthesiology, applied
Meditation therapy
Magnetic therapy
Mesmeric Vital Energy
Mind-Body therapy
Music therapy
Myotherapy
Naturopathy
Neural therapy
Nutritional therapy
Orthomolecular Medicine
Osteopathy, manipulative
Oxygen therapy
Parapsychology
Pharmaceuticals- oral , injection
Physiotherapy
Qigong
Reconstructive therapy
Reflexology
Rolfing
Self Healing
Sex therapy
Shiatsu
Surgery
T'ai Chi
TNS
Touch and hands on therapy
Trager
Verbal Charm Therapy

Visualization therapy
Waving hands over patient
Postural therapy
Prayer therapy
Psychic Energy therapy
Psychoneuroimmunology
Psychospiritual Education
Psychotherapy
Yoga

Shang was pleased with himself. Now it was time to learn what they all meant. "Dad, do you think we can give better meaning and understanding to all of these approaches to healing, now that we've heard Jonathan's Eureka?"

Dr. Shu smiled. "I think we should go back to the Creator of the Eureka to see how he offers us some understanding as to how these enormous numbers of therapeutic approaches could all have some value in the treatment of different illnesses."

Jonathan knew that if he was presenting a new theory as to the workings of Alternative Therapies, he would have to demonstrate their practical use. "Thank you Shu. I knew that before the day was over I would have to make some sense out of the Alternative therapies or the theories would be nothing but words, as are so many of the scientific theories that have come and gone since the 1600s. Before we begin, I would like to add one more thing to the list that can give us a clue as to how many of the therapies work, and that is the Placebo effect. There are many in Medicine who feel that 60% of the patients get well because they have faith in the education or the healer. I firmly believe that this is true and the polarity of the thinking of the patient may make the difference between life and death. To illustrate this point I want to tell you the story of George, who represents a high percentage of my dying patients who lived because they were not ready to die.

Sir William Osler stated that he initially had to understand the body and mind as a whole, before he had to know the disease. Knowing everything about the patient played a major role in the ultimate treatment of any disease. Since the days of Osler, there have been losses and gains in Medicine that relate to the problems we are trying to solve regarding therapy.

George was in his late 60s when he came to me with a full feeling in his right side. The history and physical clearly indicated that there was a mass in his right kidney and after several confirmatory tests we opened him up and found a very malignant tumor that had invaded the liver, lymphatics and the major vein of the abdomen, the inferior Vena Cava. If one checked the already hardened major vessels of the abdomen, the cancer could be felt above the diaphragm and only a short distance from his heart.

We removed the kidney, demassing the cancer, and closed him up. The odds were that within 6-8 weeks he would be welcoming the death that was soon to come. His wife Jean had died 6 years earlier and I told his daughter that I thought George was now soon to follow.

When I made rounds the next morning he asked me if I had removed the kidney. I answered yes.

"Can I get out of here in the next few days" he asked.

"Of course" I answered. "As soon as you're up and about and can take care of yourself, I'll send you home."

"I don't want to go home. I want to drive to Northern Wisconsin and fish with my 13 year old grandson."

"George" I explained "as soon as you're ready to travel and your wound is healed, you may go. But that's with the understanding you stop boozing. You must keep your salt low and take all of the pills I've given you." He was a severe cardiac but never, in all of our conversations, did he ask me what was wrong.

For the next 3 years I saw him frequently in heart failure and with marked distension. Always, he admitted "I didn't do what you told me to, but I know you can fix me."

We fixed him frequently, and he was always off and running again.

It was at the end of a 3 year period that he came to me and asked that I hospitalize him. He was feeling bad. It was evening, and the next morning I received some of the blood work from the lab. On making rounds he was markedly distended and very short of breath, but appeared relaxed and not depressed.

"Doc" he said "I think I'm ready. Do you mind if I join Jeanie now?"

He had never once talked about cancer and death.

"Is your pain pretty bad now George?" I asked.

"It's not too bad, but I'm tired. My grandson is getting older. I feel pretty much alone all of the time. I'm ready."

"George, my friend, you have my permission. God bless you for your wonderful Will. You've taught me much these past three years."

We shook hands, I kissed him and I left the room. He was dead in 5 minutes. This was not the first time that a dear friend wanted to say good-bye and get permission to die. I'm sure it won't be the last.

Boys, what this story demonstrates is that a physician has the power to orient the patient within a healing force-field that the patients may or may not implement for themselves. The physician must be clever enough to help the patient discover the goals that intensify the powers of living and in this way inhibit the forces of dying.

You may ask why I didn't tell George shortly after surgery that I thought he would die soon. I'm sure if I did, he would have. I believe that the art of

Therapeutics, in the more serious terminal illnesses, is more powerful than the Ethics of Informed Consent- a code that is more of a legal sham than a process of moral and ethical worth. It is certainly not of therapeutic worth. I say this because at no time did George want to know if death was immanent and I felt no compulsion to tell him what he chose not to ask. If he had asked, I would have told him, and most likely he would have followed my command and died within 6-8 weeks. It was the goal that worked the miracle. The goal George found was simple. He wanted to live until his grandson was 16. He felt that after 16 his grandson would no longer be likely to enjoy fishing with grandpa, because of a new rising interest in girls. That meant George was giving himself 3 years. He lived 3 years and that was 20 years ago.

Today we know that T lymphocytes, our powerful immune warriors, are influenced by attitude. In the last 26 years that I've practiced Medicine, I have not seen many cases of cancer that were not preceded by a crisis moment of depression.

What is the lessons we can learn from this story?

The power of the cell is in its ability to integrate its structural proteins and energy needs, so that it can carry out its various intrinsic responsibilities, and its responsibilities to the giving organ environment that surrounds and contains it.

The power of the organ is in its ability to integrate its cells and their multi-functional potentials, so that they can serve their own needs and the needs of the systemic environment that they serve.

The power of the systems, the cardiovascular, gastrointestinal, genitourinary, and musculoskeletal, is in their ability to integrate their organs, so that they function harmoniously and sequentially toward their common goals. Only in this way can they serve the human host which they have been designed to serve.

The power of the human being relates to his ability to integrate these systems, organs and cells, and harness them for the purpose of actualizing their greatest potential. Healthy and successful living requires the integration of all that of which we're made.

Toward what end?

To serve the goals associated with every relationship we've established in the many systems of operation which have captured us during a life time. To understand the relationship between cells, organs, organ systems and human beings is the wonderful and endless goal of Clinical Medicine. It is more a therapeutic art and science, with tools that are intuitive, rational, spiritual and technical.

In the past, the power of the physician was in his sustained, long term relationships with his patients and in the unequivocal trust that existed between the two. The patient was unafraid to share all and the doctor was unafraid to say

"You will get well".

Jonathan was through, and all was quiet in the room for a short moment. Dr. Shu then broke the silence. "I think Jon that you are saying that before the technology of today, the very doctor-patient relationship, essential to influence the prognosis of all illnesses, was trust, affection and a belief in the power of the physician. Having said that, what do you think are all the elements that make this possible? Certainly the same relationship is essential in Chinese Medicine and what I've learned, it is also present in Ayurvedic Medicine."

Dan was eager to respond to Shu's statement. "I think that Jon has touched on the main, universal factor that explains, in part, the success of all the Alternative Therapies listed, including the so called Placebo effect. It is the precipitation of a contracted, Para, loving relationship between the physician and patient, and since Para is the healing mode for all structure, it heightens the effect of the immune system. These are two of the mechanisms essential for healing. Is that right Jon?"

Jon had a big smile on his face as he responded. "Dan, I believe you're absolutely right, so I think it would be a good idea for you and Shang to re-examine the entire list of Alternative Care offerings and discover what part of my story about George gives insight into what magic process makes them work. I would like to search more intensely into the ancient Medicine of the Chinese culture with Dr. Shu. I think that both you and Shang should go off to another room and work on this project, so that we can accomplish, by the end of the day, as much as possible. Is that alright with you Shu?"

Shu was a little surprised at what was happening, but then realized that Jon had come with the very special purpose of getting as much information as necessary to ultimately make it possible for Dan to design the Ortho-Para Research Center. In this way it would accelerate the process. "I think that would be fine Jon. Why don't Dan and Shang go down to the basement where they'll find a library of Alternative Medicine. When you get stuck with any of the offerings of the various modes described, I think you'll solve your problems by using these books."

Shang and Dan disappeared quickly and left Jon and Shu alone with the difficult task associated with understanding Chinese Medicine in terms of Western Medicine today.

Having comfortably equated Yin-Yang to Para-Ortho, Jon wanted to examine the Five Phase Correspondences and compare them to something he could comfortably incorporate in his own mind. It would have to have, as a foundation, the structure and function he understood from a Western Medical perspective.

While Dan and Shang were going to the library, Shu brought out a pot of tea and poured it for both of them. As he poured, he spoke. "Jon, I sensed you're having a problem with some aspects of Eastern Medicine. What can I tell you?"

Jon was quick to answer. "I'm having some difficulty translating your Five Phase Correspondences into a concept that I understand. I realize we're talking about 4 thousand years ago and the workings of intuitive minds that had yet to be exposed to the partistic analyses of body structure and function that we now understand from a physiological chemistry perspective. Perhaps you can help me understand the diagnostic significance of the five categories, and in what way they express significant information regarding health and illness."

Shu quickly got up and excused himself. "I must go down to the library and bring up a chart that we can refer to while we're talking."

He returned very quickly with the chart, and began what would be an interesting and comprehensible explanation that Jon quickly began to understand.

Category	Wood	Fire	Earth	Metal	Water
Viscus	Liver	Heart	Spleen	Lungs	Kidney
Bowel	Gall bladder	Small bowel	Stomach	Large Bowel	Bladder
Season	Spring	Summer	Late Summer	Autumn	Winter
Yin-Yang	Yang in Yin	Yang in Yang	Early Yang In Yang	Yin in Yang	Yin in Yin
Time of day	Before Sun	Forenoon	Afternoon	Late Afternoon	Midnight
Climate	Wind	Heat	Damp	Dryness	Cold
Direction	East	South	Center	West	North
Development	Birth	Growth	Maturity	Withdrawal	Dormancy
Color	Cyan	Red	Yellow	White	Black
Taste	Sour	Bitter	Sweet	Pungent	Salty
Sense Organ	Eyes	Tongue	Mouth	Nose	Ears
Odor	Goatish	Scorched	Fragrant	Raw Fish	Putrid
Vocalization	Shouting	Laughing	Singing	Weeping	Sighing
Tissue	Sinews	Vessels	Flesh	Body Hair	Bones
Mind	Anger	Joy	Thought	Sorrow	Fear

You must understand that we're talking about 4000 years ago. Although it has been modified over these many years, the basic tenets that created these categories are ancient. They're the products of intuitive minds and not the analytic and scientific minds which drive our thinking today.

Let's start with the seasons. From our understanding of the relationship between the polarity of darkness and that of light, one can relate the Yang in Yang to summer, and the other variations very much in relation to the position of the sun during the year's cycle. One can understand that during the different seasons the incidence of different illnesses vary greatly.

The time of day also produces a variance in the light and darkness, and just as the seasons influence Man's potential weaknesses, so does the time of day. The climate also fits into the same category and Western Medicine also responds to the particular illnesses associated with the variable climates we find throughout the world. Ancient China was very large and it had exposure to many climatic conditions which influenced the diseases found in these different areas.

The concept of Direction is somewhat different. You must note that China is placed in the center and the rest of the world was north, south, east or west of the center. Although Marco Polo was a westerner, who became famous for his travels to the East, there were many Chinese who traveled throughout the world, so that the predilection for different illnesses and climates became very well known to the Medical taxonomists of that day. You must remember that Man looked at the sky and its uniformity of structure, and then looked at Man who lacked this uniformity. He then tried to categorize Man in relation to that which was stable and predictable, and in this way make man more predictable. Astrology became the power of the Babylonians, and complex Eastern Medicine became the birth child of the Chinese. Out of Astrology came Astronomy and out of the concepts of Chinese Medicine we find the root systems of Herbal Medicine that eventually became the base of the entire Pharmaceutical industry.

There is no country that can equate itself with China as far as the length of time it has existed as a nation. It has given the nation and its people the time to slowly mature in relation to a uniform culture, so that compared to all the surrounding world, China is by far the most mature. Remember, I'm saying mature and not necessarily modern. If we look at our country today there could be much debate as to whether the modern United States is truly as mature as we would like to believe. We are still experimenting with a Constitutional Republic, which is slowly becoming more and more Democratic, and we are only now discovering some of the problems that are born in a free society made up of a high preponderance of the needy, and a financial class system, that at this moment in history, dominates the Federal hierarchy. The relationship between maturity and modernity has not been revealed to us yet. It is still possible that we can learn important lessons from China."

Jon wanted to interrupt and discuss this point with Shu, but Shu had not finished and continued on with his description of the Five phases.

"As far as the color that was designated in the Earth column, you must remember that the Chinese are the yellow race and everything that was superlative was placed in the earth Column. This information was privy to all the Chinese population, and it was considered politically correct to place China and all of the positives of life in the same column. You'll note that the most conservative and mindful brain function, thought, was also put in the Earth Column.

As far as the organs are concerned, the solid organs were placed in the viscus column and the hollow organs were placed in the bowel column. This, I'm sure, is related to the other overt values that were more easily seen in the outside world and were, over a period of time, related to the organs and the diseases more prevalent at different periods of time. I can assume that there was an empirical base to those classifications and may very well have been changed during the different dynasties. I think that covers the five phases but I should add one more very important part of Chinese Medicine and that is the concept of Qi (Chi). Chi, in this country has always been equated to Energy. Accepting this as a given, it has been divided into 5 types, Ying Qi- which supports and nourishes the body; Wei Qi- which protects and warms the body; Jing Qi- which flows in the channels of the body; Zang Qi- which flows in the organs of the body; and Zong Qi- which is responsible for respiration and circulation. How, Jon, would you translate that into a comprehensible Western equivalent of energy?"

Jon had sat attentively during the entire exposition. Some of the confusing concepts were now clear and he enjoyed the interpretation suggested by Shu because he hadn't related the mental evolvement of this ancient culture to that of the Babylonians which he understood more clearly.

"Dr. Shu, I want to thank you for your dissertation. It cleared up many points for me, especially when I realized the antiquity of the culture that the data was coming from. In answer to your question as to the different types of Qi, I might suggest the following. Construction Qi is the Potential Energy of structure. Defense Qi is Kinetic Energy at its source. Channel Qi is the flow of Kinetic Energy via the Autonomic nerves and in the walls of the lymphatic and vascular channels. Organ Qi is the energy that controls the intrinsic function of all the organs. Ancestral Qi refers to the energy that intrinsically controls the automatic reflexes, controlling respiration and circulation, and probably some of the other reflexes present at birth which are deconditioned because they're never reinforced by use. You know Shu, this is the first time I'm aware of the pathways that are theoretically possible for Acupuncture and Moxibustion therapy."

Shu smiled and was very pleased that the translatory process was slowly taking place. "Maybe you can help me go a bit farther with the translation. In my mind I have two systems of thought that I never mix when I'm making medical decisions. The translatory process is too difficult for me. How, in your mind, does the autonomic nervous system come to terms with these five types of energy and the ability of those of us who do Acupuncture and even Acupressure, to achieve the results we do achieve using traditional Eastern Medical techniques?"

Jon listened carefully and presented a questioning face to Shu. "That's a puz-zlement," he said with a smile "but let's go back in our thinking to the origins of the Autonomic Nervous System."

We must remember that as we look back at the Phylogenetic recapitulation that Ontogeny represents, at one time, up to the period when Lampreys were the most complicated creatures on earth, the only autonomic system that was dissectable was the Parasympathetic. Those creatures that pre-existed the development of the Orthosympathetic Nervous System, did not have the focused chemistry that coordinated the means of coping with adversity, so that their survival as a species depended more on their number of progeny and their reproductive excesses, than on their ability to sustain themselves in an adver-sarial environment.

To go back a little further, we note that the Orthosympathetic Nervous System arose when the metacellular creature required a circulatory system capable of transporting the sources of life energy long distances from their environmental source. When the environment was unable to succor the needs of the cells of a passive living creature, active, mobile living creatures came into existence. When active, mobile creatures, within the environment, could not provide nourishment to all of the cells of their inner environment, the cir-culatory system arose and with it the controlling Chromaffin cells, and later the Orthosympathetic cells. Thus, even as we look back at our phylogenetic evolvement, we see evidences that the resistances of the environment, that frustrate the attempt of the organism to serve its survival needs, precipitate the changes essential to its ultimate survival. These changes involve the movement of the environment within the organism, controlled by the Autonomic Nervous System, and the movement of the total organism within an outer environment, controlled by the Central Nervous System. The power of the Orthosympathetic Nervous System grows progressively with the increase in the outward focus-ing of the organism. The flow of Perceptions, of Ideas and Concepts are the final by-products of Man's active involvement with the outside world. It rep-resents the highest level of Autonomic and Central Nervous System evolve-ment and fusion.

This Evolvement Study was performed by Walter Gaskell in the early part of the 20th Century. He noted that in the worm, the dominant cell mass of the Sympathetic Nervous System was the Chromaffin cell. As we move up the Phyla, from invertebrate to vertebrate, we see the development of the sympa-thetic ganglion cell which arises from the same cell mass embryologically as the Chromaffin cell. Both are associated with the development of vascular musculature and in those creatures that have no vascular musculature, such as the Amphioxus, there are no chromaffin or sympathetic cells to be found. In the lowest vertebrates, the cyclostomata or early fish, there are no sympathet-ic ganglionic cells and only masses of Chromaffin cells that can be found along the segmental cardinal veins. Without going into any further detail, the most primitive creatures have evidence of a vagal nerve that controls Parasympathetic enteric functions. To repeat- the most primitive position of the Autonomic Nervous System is Vagal and therefore Para. It is interconnected with intestinal function and is the controlling force for all intestinal motility. With the further complexification of the vascular system and the need for isolated and specialized control of different portions of the vascular musculature, there is a slow transition from Chromaffin cells to Sympathetic cells, with the associated development of the stress backup of the mass of Chromaffin cells in the medulla of the suprarenal gland. The Pheochromocytoma is therefore a tumor of the primitive Chromaffin cells and presents as a vasculature muscular problem, Hypertension.

Please bare with me for a little bit longer. Since it is apparent that the Para Nervous System is older and preexists the Ortho Nervous System, it is interesting to see the different placements of the ganglia of the Para, which are closer to each organ and are uniquely functioning with that organ, while the Ortho ganglia are located farther away from the organs and are attached to not one, but all of them. This suggests that not only does the Adrenal gland influence all of the organs of the body via an Adrenalin surge, but the Ortho ganglia are also generalists in relation to outside activities and influence all of the body's functions. Ortho activates the body in relation to outside activities and makes available that energy necessary to perform those functions. Para protects each organ, and dependent on the particular needs of that organ, it performs its regenerative or reconstructive functions when Ortho dominance disappears and the Para polarity can dominate and control body chemistry."

Jon stopped for a moment and Shu saw his chance to jump in. "I see now what you're implying. There is a way, via the vascular and lymphatic network, to attach the cutaneous structures to the internal structures they were designed to protect. Certainly, in medicine, we understand the radiation of pain from different organs to areas of the body external and very distant from the source of pain. What reason can we have to only allow the transfer of pain outward and not allow the transfer of external stimuli inward? Certainly that is what Acupuncture and Moxibustion represent."

Jon was aware that the conversation was slowly getting more complicated and that some of his implications were still not clear. It was very apparent that the Parasympathetic Nervous System was the more ancient of the two components of the Autonomic System, and that Ortho was clearly dependent on the needs the creature had in relation to the environment in which he lived. Another thing was becoming clear as Jon directed his questioning analysis to Shu. "Do you understand that we must recognize the antiquity of the Parasympathetic polarity and that organisms tend to be in Para dominance unless there is an interaction with the outside world? That would mean that

Ortho dominance can be as frequently caused by Para suppression. The determination as to dominance, may not mean the polarity that is dominant must be suppressed. On the contrary, the polarity that is suppressed must instead be elevated to suppress the dominance of its antithetic antagonist. Am I making myself clear?"

Shu smiled as he answered. "I think I know what you're saying and it is understandably very significant. If one has an Ortho dominant illness and we try to create a Para environment to find the essential autonomic harmony and balance, would it be harmful if the Ortho dominance were actually due to a Para suppression? Would it actually be harmful if the Ortho dominance was due to Ortho excesses? Would we try to balance it with Para therapy? If the Ortho dominance is due to Para suppression, shouldn't we do exactly the same thing as we would if Ortho dominance was due to a true excess, because in both cases we are trying to balance out the Para with the Ortho and thus create the harmony essential to good function? I believe that either dominance, whatever the cause, must be treated by instituting the creation of the appropriate antagonist to find that point of balanced antagonism, homeostasis, and functional harmony that represents good health and a healing environment."

Jon was embarrassed. Shu was absolutely right. He was thinking in terms of the Yin-Yang harmony he had been taught in Chinese Medicine and Jon had to get as comfortable with his Ortho-Para theory as was Shu with his Eastern Yin-Yang. For the first time Jon realized that it was one thing to develop a theory of cause and another to actually begin using this theoretical structure in a functional capacity. "Shu, I'm so happy we've had this session. I would be much more comfortable if you would give me a brief primer on the treatment of the typical illnesses that enter your office each day. It would get me on the right track in thinking with your Yin-Yang thought patterns."

Shu bowed his head in compliance. "Let me first refer to your first Treatise in which you speak of contraction and expansion being in a state of alternation, and that there is a 'mean' which the reciprocal forces seek to maintain, no different than the swing of a pendulum. Dependent on the phenomenon involved, the interplay of Yin and Yang shows in various ways: active and passive; overt and covert; expansive and contractive; radiant and concentrated; hot and cold; and night and day, are all mutual antagonists that seek harmony as they flow with their basic oscillations.

A good example that attaches itself to people would be the way Yin and Yang are expressed in the observations of gender. Women are said to be Yin on the outside, soft and yielding, but Yang on the inside, firm and resilient, while men are Yang on the outside but Yin on the inside. As Daniel Reid described in his book on "Chinese Health and Healing", there are even women who are more assertive and Yang than men, and this type of woman tends to gravitate towards a man who is more yielding and Yin than others. Similarly, extremely macho men tend to prefer more docile, yielding women. Nothing has only one polarity, and everything tends to seek a complementary opposite that can be felt as balance."

Jon asked "Could you give me an example of what you would do if a patient came in with a specific condition that you interpreted in an Eastern way?"

Shu responded "Certainly. If you read the excellent book by Daniel Reid you would see examples of what I see each day. What is wonderful about this therapy is that you don't have to go to Medical School. You do not have to be a doctor to practice the wisdoms that simple peasants have practiced for several millennia. When we're healthy and environmental conditions are normal, our bodies adjust automatically to the cyclic changes of seasons, days, months and the weather. However, when human vitality is weak, from stagnation or illness, or when environmental changes are extreme, the human system can fail. From the Chinese Internal Medicine Classic we hear stated, "Physicians who excel in diagnosis, first check the complexion and pulse. They then analyze the condition in terms of Yin and Yang. Just as maintenance of optimum Yin-Yang balance is the key to health, correct analysis of Yin-Yang imbalance is the first step in diagnosis of disease. Symptoms of excess Yang for example include fever, sweating constipation, chronic thirst, dry lips and mouth, dark urine, heavy breathing, rapid pulse and irritability. Excess Yin is reflected in such symptoms as chills, cold hands and feet, loose bowels, lack of thirst, shallow breathing, slow pulse and lethargy. Chinese therapy focuses not on the temporary relief of these various external symptoms as modern medicine does, but rather on correcting the basic internal imbalance of Yin and Yang which causes the external symptoms. When the root cause is corrected, all symptoms disappear naturally. This is called 'treating the root, not the symptoms'".

The Chinese Internal Medical Classic also says "If it is hot, cool it down. If it is cold, warm it up. Therefore cooling Yin herbs, such as rhubarb and Senna are prescribed for acute constipation, which is a hot, Yang condition. Warming Yang herbs such as cayenne and ginger are given for such cold, Yin conditions as colds, flu, poor circulation and indigestion. It is the doctor's responsibility to learn the various, specific herbs that more effectively treat specific symptoms."

Jon had been listening attentively. "Then, if I properly equate Yin with Para, and Yang with Ortho, I should be able to comfortably choose my herb pharmacopoeia using the same diagnostic approach as you. I'm absolutely amazed that I've been ignoring this aspect of Medicine for all of these years. I know that the Taoists use the same logic to design their lives and their concept of balance extends itself to all areas of their lives.

It's nearly time to ask the boys to join us, but before that I want to make a statement that may influence our analyses in the future. As we stated before, when life began the creatures were initially Para with the Ortho movement

71

aspects of life being created by outside forces, probably ocean currents. When the heart and vascular system came into being, the Chromaffin cells that formed humoral Ortho substances, became a reality. As the complexities of the invertebrates became the complexities of vertebrates, the Orthosympathetic cells became embryologic realities, and when thought and speech evolved, the Central Nervous System was formed, all being sequentially created when the complexities of life demanded it.

When a human egg is fertilized, it goes about recapitulating the phylogenetic evolvement process so that all fetuses go through the same phases of evolvement that the creatures of the past went through. When a fetus becomes an infant, it too must go through the sequential phases of social maturation, not dissimilar to the embryos journey when in utero. I wanted to say this because it's essential to the understanding of some of the consequences. When illness creates a reversion in perceptual sensitivities, and with that reversion, an involution to different levels of Autonomic polarity, they are more global and protopathic in their pickup than the mature, epicritic sensitivities of an Ortho or Cerebral Dominant mature adult.

Shu, I don't want to stop now, but I feel it's time to ask Shang and Dan to join us and see what they've accomplished with the Alternative Care chore we gave them."

Dr. Shu rose to his feet. "I think you're right Jon. I'll go downstairs and bring them up so that they can share what they've accomplished in the last two hours."

Jonathan was thrilled with what he had learned in so short a period of time. It suddenly didn't feel that the time he spent in learning other Complimentary Medical procedures would be a chore. His admiration for the Chinese intellect was slowly growing and he was curious as to what the other Alternatives had to teach him.

Chapter 7

When Dan and Shang returned to the conversation, they were both elated over what they had accomplished in the subterranean basement. There were many books on the various Alternative approaches to Healing. Although Dan was less familiar with them than Shang, the concepts about which system analysis was based was that which Dan had learned from reading Jon's Treatises.

Dan enthusiastically began the conversation. "In order to classify the listed Alternative procedures, I suggested the use of the Embryonal classification of the diseases being treated, so that we could quickly understand what we were treating in way of its Ortho and Para consequences. To do this, we first listed the Embryonal dominances and applied a theoretical Ortho/Para ratio to each of the Ages within the Eras already described in the Treatises. The following demonstrates exactly how the ratio assisted us in giving a good idea as to what disease dominance prevailed.

The Endophilic Era was divided into three Ages- The Endophilic Age with a Para/Ortho ratio of 9/1; Mesoendophilic Age with a ratio of 8/2; and Ectoendophilic Age with a ratio of 7/3. The Mesophilic Era was divided into three Ages. The Endomesophilic Age had a ratio of 6/4; the Mesophilic Age with a ratio of 5/5; the Ectomesophilic Age with a ratio of 4/6. The Ectophilic Era was divided into three Ages also. The Endoectophilic Age, with a ratio of 3/7; the Mesoectophilic Age with a ratio of 2/8 and the Ectophilic Age with a ratio of 1/9.

The ratios clearly represented what the potential dominances of the diseases were, and therefore what therapies would ultimately be needed to counteract the dominant disease polarity with which we were working.

(A) The depressions are markedly Para or Yin dominant. They would fit into the 9/1 - 7/3 ratios we've just described. They required a therapy that would add a powerful Ortho or Yang component to their life. This could be accomplished with several different modes. Aerobic therapy, goal oriented therapy, Acupuncture therapy, Moxibustion therapy and Specific Herbal stimulants. Western Medicine offered Serotonin reuptake inhibitors, Tricyclic Antidepressants, stimulants in the Dexadrine family and several others whose exact actions are not understood. Considering the side effects of the Pharmaceuticals, the goal oriented therapy is certainly the road to a sustained improvement without the potential toxicity and addiction that drugs represent.

(B) Severe and Hyper-manic states were Ortho and would have a ratio in the neighborhood of 3/7-1/9. They required a therapy that would add a powerful Para or Yin component to their lives. This type of illness could also be approached in several different ways. The Alternative or Eastern approach would be Meditation, Aroma therapy, Touch therapy, Love therapy, Water therapy, Music therapy, Herbal and Flower therapy. The Western approach would be Sedation, Anti-manic agents, and Tranquilizers.

(C) In the classification of situational Anxiety states, which may be the most common emotional problems faced in the Western world, we've suggested the ratio was 6/4- 4/6. The Alternative or Eastern approach would be ritualized and daily exercises, Qijong, T'ai Chi, Aikido, any ritualized and repeated activity that is of worth. Any Goal Oriented activity is a wonderful method of correcting the polarity of either the Ortho or Para excesses, especially if the goal is considered worthy by the patient.

Having divided the Emotional Illnesses into three main Embryonal categories we felt we could more easily classify the different therapies, all of which would certainly help some people or they would not exist today. Certainly some of the therapies can be due to the relationship between the physician and his patient."

Shang was ready to step in, and as soon as Dan stopped talking he began his discussion on a different aspect of therapeutic magic. "I want to refer back to your description of the perceptual journey back from the Ortho polarity of a distant relationship, to the Para polarity of intimacy. If there is any universal multi-polar relationship, it is that which can be initiated by a caring physician. When a true Para relationship has been established between the doctor and the patient, healing begins the moment they see each other. Dependent on what is being shared, and the inceptual and perceptual search initiated by the physician, he can create an environment of trust and understanding that totally changes the polarity of both physician and patient. I emphasize the essential nature of knowing just how far the intimacy can go if the professional relationship is to continue to effectively influence both of their lives. Dr. Jonathan, is that what you experienced in your practice, since I understand you went through a period when you were delivering babies?" Jon was quick to answer. "Everything you've said is true Shang. For the patient, and rarely for the doctor, the closeness occasionally gets dangerously close to love. The doctor must be aware of it and discourage it when it surfaces. The doctor-patient relationship is most therapeutically effective when it allows for a professional intimacy. The alternative is to maintain a distant relationship with all patients. The physician would then have a minimal therapeutic effect on the patient and would have to use drugs instead of the psychologic skills that might have avoided their use. I might add that along with your suggestions regarding the effect of the therapies, you've already listed for the various Para/Ortho ratios, the most effective is a long, time consuming visit, similar to what the patients now have with most Alternative therapists. The very thing, that the doctors used to offer their patients in closeness are now offered by the intimate Alternative therapists. Many times I've written about the Tech and Touch changes that have taken place in Medicine in the last 50 years. Only when the Touch returns to Medicine will the physician once again be fulfilling his role as a teacher and healer to those who seek his help."

Shu happily listened to Jon who had rapidly revealed that his concept of a Physician was the same as his. "Thank you Jon for emphasizing the relational importance to the doctor-patient relationship. The very techniques used for diagnosis in Traditional Chinese Medicine encourage the very tactile relationship you're suggesting. I think that is what influenced Rev Rebecca in establishing her relationship with me shortly after she accepted her pastoral role in the church. Well Dan, do you want to create a list of all the Alternative Therapies and suggest in what way they might work, or toss them off as irrelevant phonies?"

"That would be fine Dr. Shu, but understand, if the therapeutic response is more due to the relationship than the other mental or physical therapies offered, we could be seeing positive results with all therapies and they would be impossible to differentiate from the placebo effect. These healers would have accomplished what the patient wanted and probably at a cheaper price. What is certain is the uncertainty regarding the actual reason for therapeutic success. Are you ready Shang to help me respond to your dad's request?"

Shang laughed, as he responded. "I'd better be. Shall we just use the alphabetical order we started with, and designate therapies? I know that a great number of the problems, that now respond to surgery and antibiotics, are the responsibilities of the physicians who practice the traditional Western Medicine we teach in this country.

Dan looked questioningly at him. "Let's make it a little easier. The therapies that you designated for Para illness, we'll refer to as (A). The therapies that you designated for Ortho Illness, we'll refer to as (B). The therapies that you designated for Ortho-Para Illness, we'll refer to as (C). We'll understand that an intimate relationship is beneficial in all illnesses, but is not necessarily a substitute for any. We then can always refer to the lists, alphabetically designated, and move on from there as we continue our studies."

Shang was in complete agreement and the following list was generated.

Therapy	Class of Illness	History of Success
Acupressure	(A)	Yes
Acupuncture	(A)	Yes
Aerobic Exercises	(A) (C)	Yes
Aikido	(C)	Yes
Alexander Technique	(C)	Yes/No?
Aromatherapy	(B)	Yes/No?
Art Therapy	(A) (C)	Yes/No?
Aston Patterning	(C)	Yes/No?
Bioenergetics	(C)	Yes
Biofeedback	(A) (C)	Yes/No?
Biomedicine	(A) (B) (C)	Yes
Body Manipulation	(C)	Yes
Cell Therapy	(A) (B) (C)	No/Yes?
Chanting Therapy	(A) (B) (C)	No/yes?
Charismatics		No/Yes?
Chelation	(A) (B) (C)	Yes/No?
Chiropracty	(C)	Yes
Cold Laser Therapy		No/Yes?
Color Therapy	(A) (B) (C)	Yes/No?
Colon Therapy	(A) (B)	No/Yes?
Dance Therapy	(C)	Yes
Detoxification		Yes
Dietary Medicine	(A) (B) (C)	Yes
Divine Healing	(A) (B) (C)	Yes/No?
Drumming	(A) (B) (C)	Yes/No?
Electromagnetic Therapy	(B) (C)	Yes/No?
Environmental Therapy	(A) (B) (C)	Yes/No?
Enzyme Therapy		Yes
Exercise Therapy	(C)	Yes
Feldenkrais	(C)	Yes/No?
Food Therapy	(A) (B) (C)	Yes/No?
Flower Therapy	(B)	Yes/No?
Guided Imagery	(B)	Yes/No?

Heat therapy	(A)	Yes
Hellerwork	(A) (B) (C)	Yes
Herbal Therapy	(A) (B) (C)	Yes
Holistic Therapy	(A) (B) (C	Yes/No?
Homeopathy	(A) (B) (C)	Yes/No?
Humor Medicine	(A) (B) (C)	Yes/No?
Hydrotherapy	(A) (B) (C)	Yes
Hyperthermic Therapy	(B)	Yes
Hypnosis and Imagery	(A)	Yes
Hypothermic Therapy	(B)	Yes
Juice Therapy	(C)	No/Yes?
Light Therapy	(A)	Yes
Love Therapy	(A) (B)	Yes
Kinesthesiology	(B)	Yes
Meditation	(B) (A)	Yes
Magnetic Therapy	(A) (B)	Yes
Mesmeric Vital Energy	(A)	No/Yes?
Mind-Body Therapy	(A) (B)	Yes
Music Therapy	B)	Yes
Myotherapy	(B) (C)	Yes
Naturopathy	(A) (B) (C)	Yes
Neural Therapy	(B)	Yes
Nutritional Therapy	(C)	Yes
Orthomolecular Medicine	(B)	Yes
Osteopathic Manipulation	(B)	Yes
Oxygen Therapy	(A) (B)	Yes
Parapsychology	(C)	Yes
Pharmaceutical- oral or injection (B)		Yes
Physiotherapy	(B)	Yes
Postural Therapy	(C)	Yes
Prayer Therapy	(A)	Yes
Psychic Therapy	(A) (B) (C)	Yes
Psychoneuroimmunology	(A) (B) (C)	Yes
Psychospiritual Education	(A) (B) (C)	Yes
Psychology	A) (B) (C)	Yes
Qijong	(C)	Yes
Reconstruction therapy	(B)	Yes
Reflexology	(A) (B)	Yes
Rolfing	(A) (B)	Yes
Self Healing	(A) (B)	Yes
Sex Therapy	(B)	Yes
Shiatsu	(B) (A)	Yes

Surgery	(B)	Yes
T'ai Chi	(B)	Yes
Transcutaneous		Yes
Nerve stimulation	(B)	Yes
Touch or Hand Therapy	(B)	Yes
Trager		No/Yes?
Verbal Charm Therapy	(A) (B) (C)	No/Yes?
Visualization Therapy	(A) (B)	No/Yes?
Waving of the Hands	(A)	No/Yes?
Yoga	(C)	Yes

(Note:The Author would like to make it clear that he has varying degrees of familiarity with the various Alternative Healing Arts.)

Jon responded positively to the work of Shang and Dan. "I want to thank you both for the wonderful summary that I'm sure we'll use as we continue our study into the magic of some of these therapies."

Jon stood up and approached Shu. "I think it's time for us to go back to Rebecca and spend some time with her. I'd like permission to return to you tomorrow to focus in a little more on Traditional Chinese Medicine. You've helped me tremendously in correlating this Ancient Philosophy, Religion and Medical Art with my own theories and I'd like to pursue it a little more. I'm sure that Dan got all of the information he needed to begin the functional analysis that must precede his structural designing of this new Research Institute."

Shu was now already standing. "Jon, it was wonderful meeting you, and I want to thank you for clarifying some of the questions I had when I read your Treatises. Tomorrow, at the same time, would be just fine with me."

Dan and Shang were off in the corner, having already initiated a friendship that had lasting value. They were both in the same field and now Dan had a new colleague to whom he'd be able to talk regarding his major creative challenges, for the years to come.

Chapter 8

At 10:00 o'clock sharp Jon rang the door bell to Shu's home. Shu quickly invited him into the library. They were both enjoying good feelings in relation to that which had already been discussed in the long session the day before. They both felt that although the door was now open to an exciting fusion of Eastern and Western philosophy, translating that into a practical approach to health and illness would be a difficult goal to accomplish.

The library would be the theatre where the challenges to Eastern-Western fusion would be met by these two men who were in the most creative period of their lives. They had reached the borders between the known and that which lay beyond, and they were ready to take that first step, unaware of all of the intellectual obstacles that lay ahead. After the formal greeting, Shu brought in a pot of Herbal tea and smiling asked "Jon, have you ever drank Herbal Ephedra?"

Jon laughed "No Shu. I have not even tried Ephedrine in its pharmaceutical form. I assume your tea will perk me up if our struggle with the subject of our search today fails to give me the Yang boost I know it can. You know, I entered Medicine with a basic prejudice which may have hurt me in my daily practice. Assuming that my moments of low energy were due to systemic boredom, I would immediately enter a system that was physically or mentally challenging and in this way I escaped my Para energy prison and found the energized feelings I sought. If I became super-charged I knew that my efforts were danger-ously positive refective and I would then contract into a comfortable, more intimate system and would find my Yin or Para balance, so that the hyperki-netic state would quickly come to an end. What I'm saying Shu is that instead of titrating my feelings with drugs, and herbs are as much a source of drugs as are the pharmaceuticals, I chose to create a panoply of systems that I could enter and leave and in this way prevent my emotional pendulum from swing-ing too far to the Yin or Yang."

Shu quickly responded "That is a wonderful accomplishment Jon, but it requires a very multi-systemic life and I don't think most people have the ability to create so variegated a life that they don't occasionally need subtle therapies from the outside to tide them over a Yin or Yang crisis."

"I guess you're right" Jon answered. "The coffee, tea and cokes I occasionally drink, are proof of your contention. The food I eat has the potential for creating moments of great lethargy, especially if associated with a small glass of wine. Our culture and the tremendous availability of foods and drink, with unknown systemic effects on energy, nullifies my contention that titration can occur purely via systemic routes. I now see it as a pretty foolish statement. I suppose it shows just one of the difficult unknowns we face when we try to understand human growth, maturation and the ultimate actualization of a person's potential. Our patients, who frequently feel their doctor can make anything happen, have created a fantasy world where we are supposed to be magic heroes. That's why patients find themselves frequently going from doctor to doctor in search of a Magi. I do not believe this occurs if a proper relationship of trust is established between the two at their initial meeting. Well Shu, where do you want to begin."

Shu was quick to respond. "I thought we might discuss the ancient concept of Meridians. Assuming that Eastern Medicine's history of trial and error has found external Meridians that radiate Chi impulses to specific viscera, it would have to be by a pathway that Modern Western Medicine and Science has yet to discover. I would vouch for its effectiveness. I find it difficult to explain the effect away by putting it in the class of Placebos. I feel this for several reasons.

First, the patients most likely to have a Placebo effect are those who have not matured beyond your Endophilic Ages and maybe also in the early Endomesophilic Age. They are grown up Egocentric children who are usually dependent people, who are responsive to verbal suggestion.

Second, I have gotten excellent results when treating mature University Professors who had looked with contempt at the notion that pricking their wrist would get rid of a condition in one of their viscera. These people would most likely be immune to the Placebo effect.

Third, there are those who you would call powerful Mesophiles, who do not trust Acupuncture Therapists and who come because they've been pushed by a parent or friend. They are not psychologically inclined toward responding to suggestion and they certainly don't fit into the category of someone who I might get close enough to, to move them from their eternal Ortho dominance to a more suggestive dominance in which the Placebo effect would more likely work.

I'm telling you Jonathan that I've effectively treated all types of personality types with good results. There must be a reason and if the Meridian concept is not the answer, what is?" "Shu, I agree with you. If we can adequately explain this magic, we would have successfully married the Yin and Yang of Medical theory, which is what Eastern and Ortho-Para Medicine represents. We've pretty well established that the Meridians do not follow the branches and ramifications of the Central Nervous System. When I first began reading the literature on Acupuncture during my residency, I labored over a way to find a structural pathway that could explain these theoretic Meridia and I eventually gave up in frustration. The Central Nervous System simply had no way of conducting pin pricks to a distant, internal viscus. I equated the success of acupuncture with some form of hypnotic suggestion and let it go at that.

Time and again I was faced with the evidence that my failure to explain a truth was not going to disappear. A distinguished professor, whom I admired a great deal, Dr. Bernard Lowen, who had achieved greatness in the field of Cardiology, wrote a book on 'The Art Of Healing'. He related that on a trip to China, to which he was sent with other Specialists in countless fields, he fell and suffered a severe radiating disc compression. He was not able to walk. His fellow Orthopedic specialists hospitalized him but despite their attempts, the pain persisted. After several days of trying the American way, he decided, since he was in China, to allow a Traditional Chinese doctor to try his own special magic.

He had been visiting a small Chinese village when this happened. A very old Chinese doctor was called into service. He asked Bernard to pull down his pants and after searching for a covert Meridian, he pricked him with one of his acu-needles and asked if he felt any different. Bernard said 'No', so he moved the needle only slightly and tried again. The location was apparently right because after a moment of discomfort his pain was markedly alleviated and he was able to walk and function again. Just reading this story convinced me I had to one day explain the mechanism of what appeared to me to be an Art and not a Science. I say that because Science has an explanation, and this certainly did not. That is the position I was in when we had our first meeting yesterday."

Shu looked at Jon questioningly. "Jon, does this mean that sometime in the last 24 hours you've had another Eureka? If so I want to hear it so that we can both celebrate this moment of awakening." Shu looked elated, as he approached Jon.

Jon gave him a knowing smile and began his dissertation. "Shu, if you remember, I ended yesterday's meeting with a small description of the embryologic unfolding of the nervous system. It was while I was struggling with this step by step Phylogenetic recapitulation of Ontogeny that an idea suddenly surfaced that I took to bed with me last night. When I awoke this morning I asked Becky if she had access to an Anatomy Atlas and with a big smile she walked into the library of the parsonage and returned with a Cunningham's Anatomy Atlas. With a magnifying glass I traced the branches of the Parasympathetic nerves to the skin and realized they had their origin from the same Vagal nerve that innervated all of the abdominal viscera. The primitive nervous system, that preceded the evolvement of the CNS and Orthosympathetic nervous system, was still therapeutically functional for the primitive creatures of long ago. These ancient creatures only had a Parasympathetic nervous system and since it still existed in the more advanced animals, inclusive of man, I suddenly realized Acupuncture was no longer a miracle, but an understandable phenomenon that worked under special circumstances. If the individual was Yin or Para, which was a time that the Parasympathetic protopathic sensitivities were at their greatest, miracles could be performed that were not possible via Orthosympathetic or Central Nervous System pathways.

Shu, when I first met you, you asked me what I was doing when the Ortho-Para eureka was born. I told you that I was reading a structural psychology book that described an infant who had just finished nursing and was lying in its mother's arms in a position of flushed, eupeptic pleasure. Having delivered many babies, I can tell you that the skin manifestation of Para dominance is present at birth and I presume it is a functional power until senescence sets in. Even though Anatomists cannot perform the micro-dissections that would reveal the connection between the viscera and the skin, we know that the Parasympathetic Cranio-sacral outflow tracts innervate the skin and the internal viscera."

Jon stopped talking and watched the radiant expressions of understanding on Shu's face. Shu saw the opening and quickly responded. "Jon, that was wonderful. It also explains why the piercing of the skin can at times be very comfortable and why some people get addicted to acupuncture and return every day for treatments. When any man or woman is in a powerful Para dominant state, epicritic sensations slowly disappear and are replaced with protopathic sensations. A man or a woman engaged in sexual activity rapidly develops a peripheral sexual flush, with sensations that are initially pleasurable but then grow in intensity into an overwhelming ecstasy. Just as the Para newborn feels little pain because of his Para dominance when we do a circumcision, the Para adult tolerates the violent vibrations of the penile-vaginal contact that opens the gates to those high moments of pre-orgasmic pleasure. For that matter Viagra, the new medicine that prepares the genitals of both males and females for the pleasures of intimacy, also cause a Para flush so that more than the genitals are prepared for those moments of a pleasurable, painless existence."

Shu was exhilarated by this conversation. Jon had found the answer to both their searching minds, and Eastern and Western physicians would slowly find the commonality that would make the fusion of their Art and Science a reality. Jon was relaxed. He could see, for the first time, the answer to many other questions that had bothered him. "You know Shu, I'm feeling that many other questions will be more comfortably answered now that we've opened the evolving door. You might even call it a revolving door, as we oscillate between the Para contractions that potentiate basal brain and right brain function, and Ortho expansions which potentiate neocortical and left brain function. We use the revolving door when we go outside to our Ortho world during the day, and we use it again when we enter our Para, inside world during our sleeping at night."

Shu was relaxed and felt it was time that they both begin a critique of the eurekas that underlay so many of the explanations already given as to how everything works. "You know Jon, it is time for us to answer a few, possibly difficult questions. I've found that every discovery is usually faced with a touch of mystery. We've decided that the passageway from the skin to the viscera is the more primitive Parasympathetic nerves that supply the dermal skin via the arterial system. The anatomy textbooks say that the Parasympathetic nerves to the skin are efferent and therefore conduct the currents from within-out. How can we counter that argument?"

I thought of that last night and remembered that antidromic conduction has been found in the Parasympathetic nerves, but the details of the experiments were lacking. There are several reasons for me to believe it's true, but I'll still keep searching in the Embryologic literature. In the Embryonal theory of human evolvement we've noted that in the Endophilic Era, and all the stages associated with it, the human was primarily Egocentric and it was essential that he remain in that polarity. We can assume that is the case in all of the more primitive creatures from which we came. Thus, Egocentric cells eventually became egocentric multicellular organs and then egocentric multi-organ creatures. The human reflections of this can be seen in the location of the Parasympathetic ganglia. They are either near or in the organ that they supply, and as we described earlier, at one time they were not associated with the balancing polarizing of an Orthosympathetic nerve supply. For that matter it has been found that the Orthosympathetic Nervous supply can be completely removed from mammals, with only a significant change in their ability to work in stress environments, but they do not die. This does not and cannot occur with the Parasympathetic. One can surmise that when life was the size of cells in their watery environment, they needed no nervous system, and only the anabolic and catabolic magic associated with the oscillating relationship between Potential and Kinetic energy. When the multi-cellular creature evolved, the integration of functions required the coordination provided by the Parasympathetic control over both regeneration and replication. When it became necessary to integrate a large number of organs, in a mobile and self sustaining system, the Chromaffin and then the Orthosympathetic nerves came into existence. This added active movement to the first vascular system and then to a musculoskeletal system so that the scope of territorial control grew accordingly. Always, the mode was self survival and had to be. For the creation of larger systems capable of working together toward common ends, the Central Nervous System evolved and its function slowly grew in relation to the size and number of systems within which the individual was to function. When family arose as a system essential to survival, the concept of Allocentricity was born. As the number of systems essential to life also grew, Egocentricity was slowly weakened but always remained the foundation for individual function-al integrity and also for survival with longevity."

Shu had challenged Jon and his embryologic response was fascinating and very new to both of them. Shu continued "Jon, you covered a great deal in a very short period of time. I could picture the dynamic changes of the nervous system taking place before my eyes. It offers the imagination a great deal of imagery as we consider the differences between Para and Ortho dominant indi-viduals during the different Embryonal periods of their lives. It is here that your Ortho-Para Test may eventually add a huge dimension of personal knowl-edge and therefore potential therapy to the evaluations of the Medical Healer. In a sense it is a Yang-Yin test, and many of the Five Phase Correspondences could be added to the test and in that way more clearly define the polarity of the patient."

Jon listened carefully and was pleased with the suggestions that Shu made in reference to OPT. He thought that Helen would appreciate some of the new concepts that were born during this two day conference and he knew it would have to end that noon. Both Jon and Dan had promised to only be gone for two days, and it was now time to say good-bye to a new friend.

"Dr. Shu, my plane leaves in two hours. I must leave you now knowing that you'll remain a part of my thinking and future plans from now on. I don't know how to thank you for the new insights you've given me. I'll keep in close con-tact with you as our project progresses. I know that I've learned only a small part of your Traditional Medicine, but I intend to read some of the available lit-erature so that the next time we meet I should have graduated from a Freshman to a more understanding Sophomore in my efforts to find some of the treasures of Ancient Chinese Medicine."

Jon embraced Shu and as he was leaving found Shang, who had been listen-ing to the embryologic conversation from a distance. Shang was obviously thrilled with the knowledge he had gained during the mini-conference. He approached Jon. "Dr. Jonathan, I want to thank you for all you've taught me in the last two days and I'm sure you'll see a lot of me. Dan has offered to keep me informed as to the design of the new Research Center. The entire concept of Yin and Yang Medicine, that dad had been trying to teach me, has sudden-ly become fascinating. I must have made a maturation leap during your visit because dad has been hounding me for years and I wasn't ready." Shang turned to his father, who was beaming, "Dad, I promise to never avoid the complex subject of Traditional Medicine again." Shang hugged his dad, and shook the hand of Jonathan with smiles of gratitude that said it all.

Chapter 9

One month had now passed since the meeting with Dr. Shu, and Daniel was still fretting over his ignorance of the needs of the many Alternative Care healers who were already inquiring of Jonathan if there would be room for them in the new Research Center. Dan had already decided that Shang would be part of the design team and their telephone exchange was a multiple daily event.

As Dan was awakening on Friday morning the phone rang.

"Hello" he answered with an early morning, somewhat hoarse voice.

"Hello Dan, are you alright?" Becky answered, somewhat concerned over his faint response.

Dan's voice immediately became loud and clear, "Hi, Becky darling. Why so early? I'm fine, are you alright?"

"I'm fine, but I want you to pick me up at LaGuardia airport at 10:15 this morning. That's why I'm calling this early."

"You hadn't told me you were coming Becky. Is everything O.K.?" he was obviously surprised by the call.

"Everything is fine. I'm giving my new Ministerial assistant responsibility for the church and the services this weekend and decided it was time to spend some time with my honey bunch." Her voice was cheerful and its musical tones awakened sleepy Daniel who was now thrilled over Becky's surprise decision.

"Wonderful Becky, give me the Airline and Flight # and I'll be waiting for you with arms ready to embrace your beautiful, sensuous frame."

Becky could hear his smiling voice, "Remember Dan, I'm standing in the vestry and will have to subdue my yearning acquiescence to your challenge, but you won't find me resisting when I land in New York."

Dan laughed, "O.K. my darling. I'll be waiting for you. I imagine you have

85

to leave right now for O'Hare, so I won't keep you. See you in a few hours."

Dan was thrilled. For the past month, when he wasn't in conference with Jon and Helen, he was working at the drawing board or sitting on the grass looking at the construction site, trying to envision the new multi-systemic structure that would be his creative gift to his benefactors and teachers, Jonathan and Helen.

At this point, nearly two months had passed since Nathan had made available the initial moneys for the project, and to Dan's dismay he still had nothing of great worth to show for it. He had visited the Johnsonian Research Center where Jonathan had worked for so many years, but despite its ideal location near the East River, it didn't light any creative fires in Dan's imagination. Along with the design and structural frustrations he was still lacking sufficient knowledge as to the needs of all the practitioners who had already been invited by Jonathan to participate in this new Research Center. There were also the many who were requesting the privilege of being part of this new structure that was dedicated to emotional, physical and intellectual Wellness. Dan felt he wanted to give back the personal autonomy to those who had been rescued by the medical profession and had never found the pathway out of the maze where dependent victims live out the remainder of their lives.

Dan had been a healthy young man and knew that he was not thinking the same as his mentor. Jonathan had never depreciated the marvelous significance and progress of Illness Medicine. He had only sought to facilitate the concepts of Wellness Medicine, so that the belief that human autonomy and balanced Living and Loving would ultimately be the dominant paradigm that underlay an Integrated Medicine that allowed Healing and Regeneration, the Para powers within every living creation, to dominate the living process of all humanity. Jonathan knew that as long as the dominant Embryonal polarity was Endophilic that destructive behavior would overwhelm balanced behavior, and that Illness and its anxious and depressive sequelae would always be part of Integrated Medicine.

However Daniel himself had not reached that state of balanced thinking and dreamed that Ortho-Para Medicine would ultimately change the polarity of Thinking Man and would one day lengthen life and delay the processes of dying. With this enthusiasm and optimism, his dreamy thinking reached into areas of fantasy where Jonathan could no longer walk and was happy to latch onto an architectural dreamer to carry on the task of building a temple to Wellness.

Daniel had to park a considerable distance from the main terminal but the weather was beautiful and so was the spirit of this young lover who certainly chose for his bride to be, someone who represented the antithesis of his culture and religion. She was gentleness and softness. It stood in marked contrast to his hyperactive and very aggressive approach to life and to the resistances that would dare to interfere with his ambitious goals.

No less than Ortho needs his Para, and Para needs her Ortho, Daniel needed Rebecca and she him.

Dan did not have to wait long. The plane had landed by the time he reached the terminal and he had just enough time to walk to Gate #8 when the door opened and the crowd exited with looks of anticipation on their faces.

When Dan saw Becky exiting he got confused. He saw for the first time Laura making an appearance from behind a column that she was intentionally sitting behind. Suddenly behind Becky a stream of familiar faces appeared with big smiles on their faces, and it was then that Dan understood that a surprise was in the making.

Becky approached Dan and embraced him. "Are you surprised Danny? I thought that it was time for the entire class to get together and lend a hand in the solving of your creative problems. Everyone wanted to play an active role in making the Ortho-Para Research Center a reality, and when I sensed your frustrations over the phone two weeks ago, I called everyone in the class and each of their families had had some experience with the many Alternative Healers and felt they could add something to the creative magma from which will rise the new home of Wellness Medicine."

Dan stood with open-eyed amazement as the class, en masse, closed in on him and eventually embraced, adding laughter and chaos to the background noise of the New York airport. It was not long after they reached their luggage that Jonathan and Helen appeared and the grandiose greetings started all over again.

Jon shouted "I have a bus outside. As soon as you've gathered your things bring them to the bus and we'll take off for the Plaza Hotel next to Central Park, where you'll be staying for the next three days. Daniel, you can follow us in your car, or go ahead and let them know that 17 crazy kids will be there very soon." While he was talking, he was hugging each of the men and women who were now merely extensions of his family of children. Standing out of the crowd he saw Shang and quickly beckoned for him to approach. "Shang, come here. Class, we must have a ceremony before we get on the bus." He quickly grabbed a cane that Helen was holding. As Shang stood before him, and after an embrace, he placed the cane on Shang's shoulder. "Shang Su, I hereby Knight you and proclaim your membership in the class of 17. From now on you will be treated as one of our family and in return we expect only an occasional nugget of brilliance when we're in the throws of frustration as we zoom in on a problem that requires the brilliance of a hybrid Eastern-Western brain that only you add to this brilliant class."

The class had already met Shang on the plane, but they laughingly gathered around him again for Jon's and Helen's sake and embraced him once again. With the delivery of the final baggage they boarded the bus and the chatter among the class centered about each individual's professional activities since their last meeting at Bertha's funeral.

The sharing of class news made the trip back to the city fly by very quickly and Jonathan's voice soon cried above the background noise. "We're here and it's time to register. Find your rooms, make yourselves comfortable and be in the Conference Hall C at the Johnsonian Foundation at 2 P.M. sharp. Are there any questions?"

The class was quiet and since it was already 11:15, they had little time for gabbing and began the registration process.

At 2 P.M. they were all seated, facing Jonathan and Helen who stood before them on a slightly raised platform. Jon spoke. "We're all here today because of a concerned call from Becky who saw a very real problem that faced Dan, Helen and myself. Despite our years of involvement with the intricacies of Medicine, Biophysics and Illness, we failed to explore, in depth, the myriad of Healers who proclaimed that their techniques successfully treated those who were the failures of Traditional Western Medicine. I needn't remind you that it was my belief, nearly 20 years ago, that the technical invasion, that the physicians were beginning to represent, had dramatically altered the relationship between the doctor and the patient and at the same time nearly divested Medicine of the intimate touch relationship that was the ancient physician's primary therapeutic power.

In my search, and it certainly has not been an in depth search, I have found one commonality between most of these Complimentary procedures, and that is the hands on investment of time. It was Becky, who in watching Daniel and myself, recognized that unless we had more complete knowledge of the procedural details of what was going on in the sacred quarters of those professionals, we'd be unable to design optimal space for them to practice their healing skills.

I know that all of you are here because you felt that through personal and family experiences you were sure that you could participate not only in the designing of the space that was necessary for the healers we're talking about, but possibly give us insight as to how the therapy played a role in precipitating healing, or if not healing, an improved comfort that brought the patient back despite the fact that insurance might not pay the tab.

You have had at least two weeks to think about our problem. How shall we begin this conference?"

The young professionals sat quietly for a moment and there was obviously deep thought as to where one begins in the reaching for a solution to an all encompassing Integrative Medicine.

Dennis was the first to raise his hand, and it surprised Jonathan. Dennis had been one of the quiet boys who didn't intrude on the class discussions unless he was asked. Time and experience had apparently increased his confidence and he felt ready to set the direction of this conference. "Jon, when Becky called me, I was already thinking about the problems Dan would be facing in designing a Research facility whose goals were to understand and justify the wisdom and validity of so many Alternative Care Healers. The only evidence we had of their successes were casual, non scientific commentaries that lacked credibility and appeared more to be the so called Placebo effect that, for reasons I don't understand, Allopathic physicians have looked down upon.

Why, if a man has pain that analgesics don't alleviate and the pain disappears when he is treated by a Shaman who whispers mumbo-jumbo over his head, is the treatment called Quackery or a Placebo effect? This attitude has proven to be a weakness of Western Medicine and is, quite frankly, evidence of callous ignorance of what their role in Society is supposed to be. I hope I'm not offending Ken, Ally and Cassy, but I felt I had to preface what I'm going to say with an assertion that all that doctors should know about humanity has yet to be revealed, and I'd venture to say will not be revealed until our expanding Universe is no longer creating the new stars who will be the wonderful teachers of the future."

For a moment the class, along with Jonathan and Helen, applauded their wise colleague and only when they settled down did Dennis continue.

"After I received Becky's call, I put through a call to Steve and George. Although Steve's head is filled with rocks (the class roared) and the ancient graves of the life that walked the Earth beginning 3 billion years ago, and George's head is a little water logged (laughs again) with his Oceanographic studies of the deep, we are the historians of true antiquity and the study of the basic elemental base upon which life slowly unfolded and step by step created life, all the flora and fauna, and eventually primitive and then educated Man.

We are made up of layers of evolving time and even as we pass through the Ontogenetic recapitulation of Phylogeny, we always keep a part of the past which remains a vital part of the life process. George, why don't you continue."

Jonathan, Helen and the class were sitting spellbound as these students of pre-biotic science began lecturing the students of life. George smiled as he stood up and began speaking.

"Thank you Dennis for opening a subject I've studied on my own, but found little in the literature except that written by the students of Dr. Fox, a group of chemists who were called Constructionists and who were taking the basic elemental components of the earth, exposing them to water, heat, electricity and Cosmic rays and creating some of the molecules that are now considered the substrate of biology and life. Jonathan, we are layered creatures like the rest of the phyla of living creatures, and from the basic elemental layer, including the basic element of all matter, Hydrogen and its 92 elemental progeny, the basic molecule of all life water, the 20 basic amino acids, the 6 purines and pyrimidines, the basic nucleisides, the nucleitides, the countless DNA's and their fluid homes in the nucleoplasm and cytoplasm of the cell, we have just the macroscopic evidence of layered life and little awareness of how only small changes in the dynamic kinetics of each layer effect the total creature of which they are only a layered part. Steve, I think we've evolved into the layer of your expertise."

The class was all smiles and quietly awaited what would come next. It was becoming obvious as to what direction the three men were taking the class and thus far it was fascinating. Steve was ready for the task.

"I'm mindful that up to now Dennis and George were not speaking of life but only of microphysics and chemistry. Yet before the cell, the layers that would soon be endowed with life when the first cells were created were truly as important as the cell which could not be self sustaining and self replicating if the underlying functions of our early layers were not performing their job properly. The simple processes of self sustenance and self replication were now a permanent quality that the bi-cell, tri-cell and multi-cell would carry into all of the phylogenetic experiments. They ultimately divided into the flora and fauna that would continue the unfolding, evolving process we find in the tiny, gorgeous infant that represents the beginning of human life. Does the process end there? For the sake of brevity, I'm going to ask David, our genetic engineer, to continue walking on the path that Dennis introduced us to and see where it takes us."

David responded automatically and it became evident that the presentation that the class was now experiencing was carefully considered by all of the boys who made contact with each other at the request of Becky. David turned to Becky as he began speaking.

"It is uncanny how Becky, a servant of God, quickly transferred her allegiance to a lesser God by the name of Daniel." Laughter and giggling became the background noise. Becky quickly rose.

"You presumably brilliant basic scientists seemed to respond properly to the challenges that Nature imposed in your studies. Why was I wrong to direct your attention to the Architectural problems with which Danny was contending? And you David, don't you make fun of Dan." Becky's tone was challenging and David quickly backtracked.

"Becky, I apologize for making fun. Don't be so sensitive to my kidding. Everyone in this room would have been thrilled if we had been capable of responding to the creative needs essential to the building of this new temple to Ortho-Para Medicine. I might add that we all love Dan only a little less than you and what's more I'm sure that before this meeting is over Dan will get back at me with his eloquent tongue, as he has done many times in years gone by."

By now the entire class was laughing, including Becky and Dan and the momentary chastisement was quickly forgotten. David then continued.

"If one watches a yeast cell budding, you can quickly see how societies of cells come into existence. The first evidence of a more complicated and diverse form of life took place when the first life was created by the process of conjugation. It was then that two products of a complicated layered environment, whose products were as layered as the environment, began to merge and exchange the complex genetic material that has taken billions of years to come into existence. Sex was long in coming, but when it finally was born it was immediately considered a raging success. There were few who chose not to join the society devoted to this process of conjugation, with an intruder and an incorporator that was started 600 million years ago."

By now the class was in an uproar. They were witnessing the birth of Man and Civilization and, to a great degree, of Jonathan's three Treatises written so many years ago when he was still at the Johnsonian Foundation and fighting for the acceptance of his Ortho-Para concepts. David was not finished and continued on with his recitation on the evolvement story.

"As Jonathan stated so well in his Treatise #1, one of the Cosmic goals from day one was a differentiation process, so that it's not surprising, at all, that in response to environmental stress, life forms changed again and again, always carrying with them some residue or remnant of the past and the powers that the evolving past gave the present in the form of protection and survival value.

This is no better seen than in the evolvement of Ortho energy when the multi-cellular fish, like the sturgeon, required an Ortho neuroendocrine to energize not only the processes needed for movement within the sea but also the energy necessary for the processes of moving the sea within the vascular system.

From nuclear control of life via DNA and RNA; to cytoplasmic control of the cell via the intrinsic mitochondria and Ribosomes; and then ligand control via cell wall communication, Endocrines began controlling both sustenative and generative function and voila it was eventually followed by control via the Autonomic Nervous System and its oscillations between Ortho and Para. But that was not the end. Finally the magic of Central Nervous System control, via its Para Right and Ortho left hemispheres arose and fine tuned our responses to the outside world.

Now, who can deny that the very evolvement processes layered all of life and that we were the last species so honored by the layering process. In a sense each one of us reflect pre-life evolution and the ontogenetic history of every phylum of life that preceded us.

Why, you may ask, did Dennis, George and I put you through this brief primer of life? The answer must come in the form of a question. Can the effect of all the diverse Healers that are to be housed in Jonathan's new Research Center be due to the effect of different therapies on one or the many layers of Man?"

David smiled at Becky and went back to his seat. Jonathan slowly rose and approached the seated class and sat down on Becky's desk next to Daniel. The faces of the entire class were contemplative and in deep thought. For a moment Jonathan said nothing and just looked at the amazing talent, the flowers that had blossomed from the seeds he had once sowed when they were 10 years old.

Cassy rose and took the floor even before Jonathan responded.

"I thought it proper for me to begin the response to David's giant question. Having treated many of the layers so clearly presented to us, I would like to think out loud and see where it takes us.

The elemental first layer is quite simple to understand. The intracellular position of Potassium and Magnesium and the extracellular and interstitial position of Sodium and Calcium are essential to life, and its translation into the electrically controlled differentials because of the cells semi-permeable, selective membranes. These facts make intracellular negativity possible. All physicians realize the balanced homeostasis essential to life and when we perform blood and serum chemistries, we learn how effectively the levels of these ions are maintained, and when the upper and lower bounds are violated, we have serious and potentially lethal diseases that can only be treated by replacement and/or dialysis therapy.

As far as the Hydrogen and Hydronium concentration in the blood is concerned, the limits of pH that are tolerated before a serious illness is present is between 7.35 and 7.5. Below 7.35 we have an acidosis and above it an alkalosis. Both can kill and have to quickly be remedied.

I don't believe any of the Alternative Care therapies can be used in these critical alterations in elemental balance and only an aggressive therapy aimed at correcting the imbalance will be successful.

Allie, you've read the works of Candice Pert. Do you think the peptide ligands and their cellular receptors could be sensitive to any of the Alternative Medicines listed on the chart Jonathan gave us?"

Cassy felt that if those with medical training evaluated each layer, they might see evidence of a rational explanation for one or more of the Alternative therapies to work some form of magic. Allie quickly took up the challenge that Cassy threw her way.

"Thank you Cassy. I guess I'm the right person to discuss the peptide story since Candice was very thorough in her analysis of Endorphins, although her moves toward Alternative Care Therapy seemed as much an escape from Western Medicine as a move toward Alternative Care. Understand, I found Candace a brilliant and courageous peptide scientist, but not too bright in her attack on her male colleagues. Candace was fighting in a male dominated field and it was her responsibility to be a clever woman whose techniques had to be more seductive than deductive if she was going to succeed in influencing her

teachers and those who dominated the field. I'm sure she would do it differently today.

Her contentions were that peptides, similar to her isolated Endorphins, were
the molecules that controlled human inner affect, which Jonathan called the
incepts of Man, and thus underlay the emotions and eventually the behaviors
that they precipitated. Like most generalizations, there was some truth in what
she stated. Short chain amino acid molecules that were bonded were called
peptides. Long chain amino acid molecules, bonded by peptide bonds were
called proteins, so that in this sense Pert's contention that if both the underlying inner messages that precipitate behavior, and the structures that actuate the
behavior are short and long chain aminoacids, and they, in turn, were moderated by environmental stimuli, any one of the Alternative therapies that
brought a compassionate human being in contact with a needy person would
potentially have an affect on the person and possibly via this affect, there
would be an effect on the Illness being treated.

If the layering of life was an essential precursor to the slow evolving process,
Candace's assertions would suggest that the midbrain and forebrain were truly
not an essential miracle to the unfolding of the mammal and that the
Autonomic Nervous System and the Cerebral Hemispheres were merely ser-
vants to the demands of the 80 or more peptides that have now been identified.

It was obvious that Candace was a brilliant scientist and a very passionate
woman, and with that in mind I believe her Para dominance interfered with the
concept of the Darwinian sequential evolvement process that influences most
of us, when we try to understand the evolving process. Let us examine the next
evidence of evolvement that precedes the creation of the Orthosympathetic
Nervous System.

You must understand that uni-peptides were followed by bipeptides, and
then tripeptides until we eventually came to the Endocrines that are today, but
not necessarily, originally controlled by the master gland Pituitary. These
endocrines are the result of thyroid, adrenal, gonadal and parathyroid functions
which are hormonal representatives of the three Embryonal layers of the primitive germ cell during its early replication.

The Para functions of replication and regeneration were the survival processes of early life forms, and it wasn't until the multi-cellular development of an
internal sea, a vascular system that was energized by chromaffin cells, that
Ortho stimulant molecules were made available for controlling both the internal sea, that was the external environment of the cells, and the outside world
in which the entire organism was responding to the resistances the environment offered to life. By the time of the evolvement of Lampreys, the
Orthosympathetic balance to the preexistent Parasympathetic dominance,
came into existence.

Layer by layer, a preexistent control system that was not successfully coping with the obstacles and resistances presented at a specific moment in time, precipitated the evolution of a new control system, so that what was previously an ineffective Endocrine system was superceded by the Hypothalamic Para and Ortho control mechanisms in balanced opposition.

As the step by step vertebrate, ontogenetic history unfolded, the Mid-Brain Hypothalamic control mechanisms were superceded by the Cerebral hemispheres, which at first were only tiny kernels, but with the passage of time and the environmental presentations of greater and more serious resistances to survival, they slowly grew into the large Cerebral Hemispheres we see today. The preexistent Parasympathetic Nervous System controlled the unfolding of the Right Brain, and eventually the late appearing Orthosympathetic Nervous System controlled the unfolding of the Left Cerebral Hemispheres. We see this sequence as we observe the maturation processes at work. Up to the age of about 12, most children are influenced by a Para polarity and the behaviors controlled by this egocentric mode. After 12 there is a progressive increase in both the Ortho Polarity and the analytic components of Left Brain function and to some degree can be overtly seen in the progressive unfolding of more Allocentric behavior. In this sense Para represents Egocentricity and Ortho eventually represents Allocentricity.

Diverse humanity, as it exists today, now has the means of solving the problems of Passion and Reason; of Egocentricity and Allocentricity; of the Night and the Day; of Procreation and Sustenance; and at least early in our history, the responsibilities of Women and Men. This is a very brief summary of the history of the layering process that ultimately reached its temporary apex in Man.

Kenneth, do you think there is any point in discussing Man's evolvement on Earth in relation to our exploration of Alternative Healing?"

While Kenneth was in thought, Danny stood up and aggressively answered Ally's question.

"Kenneth, let me interrupt you for a moment. Ally's question touches on a most significant area of evolvement that Jonathan shared with Shang and I, and Shang's dad Dr. Su Shi, only a month ago. Jon, would you share it with us now?"

Jonathan smiled and quickly responded. "Dan, I'd like to hear from others in the class before I respond to your request. It is apparent that we are moving in the right direction and I want to hear Kenneth's response to Ally's question. Go ahead Ken. We have a long weekend and you'll have an opportunity to hear from me when we're summing up what this conference has taught us all."

If anyone of the 17 were asked which one, of all of them, was an emotional and intellectual extension of Jonathan, they all would have chosen Kenneth. Just as Jonathan walked in the path of Dr. Ben Yosif, so did Kenneth walk only a short distance behind. He had studied and absorbed the three Treatises on the

Autonomic Nervous System and would be the chosen one to continue studying the medical aspects of Ortho and Para, just as Laura was well prepared to take over Helen's responsibilities when the passing of the baton would one day take place. Kenneth stood up with a big smile on his face, turned to the class and began speaking.

"In answer to Ally's question, I answer with a profound 'Yes'.

I went to a conference on Wholistic Medicine in Stevens Point, Wisconsin many years ago. I was amazed at the number and complexity of the remedies being demonstrated by at least 50 healers, all of whom were foreign to me. The diseases they were successfully treating and helping appeared to me to be the functional illnesses that began as acute Ortho or Para extremes, and in time led to structural deterioration. When the illnesses were still in the functional phase, without secondary structural defects, we could see dramatic changes with healer therapies and even when structural damage was beginning there was improvement. When the damage had been present for a long period of time, comfort rather than healing was all that could be expected.

With this in mind we have to look at the various stages of human evolvement that Jonathan has taught us, and the Para/Ortho ratios that dominate the various Ages of the Endophilic, Mesophilic, and Ectophilic Eras. Since balanced opposition is our goal when we think about human autonomic balance, the therapeutic goals in Endophilia, which has so powerful a Para polarity, would be Ortho therapy directed toward the creation of system goals that would ultimately give them self confidence and a means of making a living. In Mesophilia, when balance is being successfully reached, the Ego-Allo struggle becomes an internal struggle which is always intensified by marriage, family and the many systemic resistances we must deal with. In Ectophilia, the challenge is in finding and designing enough Para time to combat the powerful Ortho, self destructive mode that dominates Ectophilia.

All we have to do is look at the polarity of the Healers therapy, determine the polarity of the patient and decide whether the polarity of the healer is capable of producing the necessary balance in the patient."

Kenneth smiled and turned to Jonathan. "Well Jonathan, am I stretching the benefits of autonomic balance too far? When Becky called, I reviewed some of your writings on the polar differentials that appeared to cause functional, maturational diseases and they all seemed to cry out for polar balance. I remember you telling me that you rarely saw a cancer that wasn't preceded by a prolonged period of depression- a most powerful Para disease. In contrast to that, the myocardial infarct was to be seen in Ortho driven, aggressive individuals. The therapy was different in both and might have been successfully treated during the functional, polar phases before the creative or destructive structural changes had changed the propensity into a reality."

Jonathan's smile indicated his pleasure at hearing Kenneth.

"What can I say Ken? You have rapidly summarized Ortho-Para Medicine and indicated why the final layer that had to be spoken to was the evolving social layer and its polar effects on human inceptions, perceptions, conceptions and the behavior that was precipitated in all of them. But Kenneth, we have gone astray.

We're trying to design a building capable of housing most or all of the worthy practitioners of Alternative Medicine. We want it to successfully encompass all of these practitioners without foolishly overbuilding, and to accomplish this end we're going to have to observe them at work.

It appears that Kenneth has successfully delineated the means of coping with the pathologies of layered Man with special focus on those that are precipitated by Ortho-Para extremes. But that was not the question that brought you all here today. We optimistically assume we understand the functional pathophysiology that is being treated by every Alternative healer. What is the space they need to accomplish these ends?

I have an idea. If all 17 of you make appointments with Alternative Care healers, I figure it would mean each one would have to see about 6 of them. We should easily be able to cover 100 different practitioners and with that information, we should have little trouble in designing facilities convenient and accessible to all of them. Does that sound feasible to you all?"

Jonathan scanned the group and Laura was the first to respond.

"Jonathan, I think you're right. I don't know any other route to the knowledge you're asking for. If we all participate in the data search, it shouldn't take any one of us more than a month to accomplish what you need."

Dan took the floor and walked to the dais. He seemed profoundly humbled by this gathering of classmates.

"I must first thank my dear Becky for recognizing my architectural needs and the importance of getting you all involved. I didn't realize the ultimate importance of getting you all involved in this project which has had a gestation period of 25 years and to a greater and lesser degree has involved all of us." He hesitated for a moment and then continued. "Becky darling, what have you planned for this evening?"

Becky quickly rose to her feet and gave him a peck on his cheek.

"How did you know I planned something?"

Dan smiled. "How do I know that the sun rises in the East and sets in the West? I just know."

The class was getting its first inkling of the soon to be marriage. Becky turned to face the rest of the class. "We are all going to the Metropolitan Opera tonight. I have 20 tickets to Mozart's 'Magic Flute' and we'll have to be there by 7:30, which means we have two hours between now and then."

Jonathan quickly spoke up, "Becky, why 20 tickets?"

Becky giggled. "Jonathan, I thought it was now time to invite George into the senior ranks. I spoke to him last week to be sure he'd be available to us in New York tonight and he gave me a highly energized 'Yes'. When you get to the hotel, he'll be there waiting for us."

Part 2

The Professional Growth of the Seventeen

Chapter 10

Helen awakened before her men and thought it would be wonderful for the three of them to have breakfast catered in their hotel suite. She quickly made arrangements with the kitchen and ordered sufficient food for five, knowing that before the morning was over they would need snacks for either of the two boys or some guest.

By the time breakfast arrived she was fully showered, dressed and ready to serve her family. It was the aroma of coffee and French toast that brought Jon and George smilingly into the living room, greeting Helen with hugs and kisses.

Jon turned to his son. "George, did you realize how wonderful your mother can be sometimes?"

"Dad, you're partially right. She is wonderful, according to her children, all of the time except possibly when she criticizes my attire when I'm going out with my friends."

Jon grinned "Now you know why I used the modifying 'sometimes'. Remember she makes me change my clothing more frequently than you."

Helen added "You can be sure that if you both didn't get chastised regularly, you'd both feel unloved. Now eat your breakfast and Jon, you have to be in the hotel conference room at 10:00 A.M."

Jon, with a full mouth, waited a moment before he replied. "I know Helen, but did you and Becky decide on what I was to lecture on today? I thought we'd let the class be free today. Most of them come to New York on rare occasions so why get them involved in a lecture? Helen, you must have had something in mind when you and Becky decided I must work on a weekend."

Helen appeared serious when she responded. "Jonathan, if you recall, you had a rather astonishing Eureka three weeks ago, regarding Alternative therapy, which you excitedly shared with me. If you recall, it shocked me at first, but when I gave it deep thought I realized we were talking about the therapy of the ancients and since it was so filled with conflict and occasionally outrage, I thought it would open everyone's mind to the scope of therapy that has been available in different venues all over the world."

Jonathan smiled broadly as he replied. "You realize this will be a form of shock therapy. You'll have to be sure that the conference room is somewhat isolated so that public access is not possible. I want to share this with the class but no-one else during this early phase of project development. I hope you understand why this should not be shared with the world right away." Jon had a concerned look to which Helen quickly responded.

"Jon, please don't worry. I took care of everything when we returned to the hotel last night. I felt that since we all meet so rarely, we would send them home with something unusual to talk about."

She was laughing when she finished and Jon joined in. George was in the dark and would learn a little more about his mom and dad at this early morning gathering of the class he had heard so much about during his entire life.

When Jon and Helen reached the conference area, the room was filled with the chattering class who were all curious as to what was to happen. Jon had decided he would conduct this class as he did with the first class he taught almost 25 years before.

Jon stood in front of his former students and smiled broadly. Helen was seated in back of the room with George and the class quickly quieted, all wondering why this addendum to the meeting of the day before was suddenly called. Jon began.

"I want to thank you all for allowing Becky and Helen to interfere with your shopping plans and reproducing a teaching scene Helen had never shared and always felt cheated when she couldn't sit in on our classes 25 years ago.

Today you have been asked to assume the role of students and from your present professional roles, attack your teacher mercilessly as he shares one of his recent Eurekas with you. But before I begin I must be sure that this is not offensive or embarrassing to any of you."

David was the first to respond. "Jon, I cannot conceive that any of us would give up an opportunity to add another creative miracle to our knowledge base." He turned to the class. "Am I right?"

Laura quickly responded. "Of course you're right Dave. It will be fun recreating the class for Helen and George and I'm sure we'll all learn something. When it becomes public, from our experience of the past, it will probably precipitate passionate debate from many professionals who shudder at the thought of any new ideas that attack the traditions of the accumulative knowledge of past millennia or yesteryear. Go ahead Jonathan, but just, as in days of old, expect your skeptical class to challenge you, if necessary." Laura's smile always fascinated Jon and he obediently nodded his head.

"Of course Laura, but be sure your knowledge base is sound when you challenge me."

The class applauded and Jonathan began the lecture that might later precipitate a lot of questions but not be challenged by this brilliant class.

"Needless to say that with the completion of my Ortho-Para test and the writing of extensive scripts, I turned my mind to Alternative Care Medicine and what profound wisdoms might be hidden in some of the mumbo-jumbo I've read over the past 10-15 years. Until I felt comfortable with all aspects of the many variations on a theme of Ortho and Para dominant polarities, I remained focused only on the Allopathic aspects of human health and behavior and chose not to digress too far from the fundamentals listed in the three Treatises.

Two years ago I made a major change in the direction I was going with my search and I began studying, in depth, the many Alternative Care therapies being offered by the enthusiastic students. Of the over 100 therapeutic techniques, each student firmly believing that their new healing powers were the panaceas for a broad range of Illnesses.

I wrote two essays that are relevant to my studies on human emotions and I want to share them both with you. The first gives you an awareness as to the pre-verbal templating of the concept of God.

The second points to the awakening of the perceptual sequence that occurs as we leave the haven of intra-uterine peace, pass through the birth channel and awaken into a world of infantile chaos. During this labored passage we are touched and massaged and it is within this passage place that we first experience our sense of taste and smell, and where we hear and see an outside world that will teach us the meaning of all the chaos our inside world is causing us.

You will see how these two essays can open our eyes and minds to the potential effect of all relational therapies which have been called Alternative Medicine but which must ultimately be included in a compendium of Integrative Medicine that becomes part of the knowledge base of all healers of the Body-Mind.

It involves not only our concerns for each moment, which remains the concerns of the growing child, but also the concerns regarding mortality, which touches on our concept of a higher Being and our eternal concerns regarding death and the hereafter.

Each day when you awaken and the serene tranquility of your nocturnal dominance still remains, it is the Para Veil that still envelopes you and is closing off the outside world. The visceral yearnings, such as hunger, have yet to awaken so that you've yet to sense the feelings of being born to a new day. You're still floating in the primal amnionic sea, having just added one day of evolvement to the 3 billion years of evolution, so amazingly accomplished in utero, during your 9 months of gestation.

Let's go back to those moments before you actually are delivered into this dry, noisy world. You still are maintaining yourself in the same embryonic posture you assumed when you slept last night, with head flexed, body contracted and peacefully floating within the blackness that speaks of eternal night. You have lived through an architectural miracle, designed by genetic codes that created your complex meta-cellular world with instructions to design specific functional units or systems in which the cells will become a functional reality, as the embryo becomes a fetus being readied for birth.

Within this primal sea the outside world is barely touching you by way of sound. There is first the beating rhythm of your primitive heart that rhythmically taps out the first beats that will one day influence the swaying motions of your gait and your movements when embraced in a musical dance. Then there is another rhythm that comes from afar and makes more acute your awareness of a primitive beating mechanism that is imposing itself on the still incomprehensible rhythms of your life.

Imaginative theories have presented themselves that encompass a subject too difficult to prove, but delightful to entertain. We'll discuss these theories in time, but first let's be born.

Are you ready? Are you ripe for your first journey into the outer world? Well, that was really a dumb question. Of course you're ready because you've accomplished all you possibly can in your dependent, primal, primitive sea. You have turned a complex egg, with great potential, into a soon to be born infant, ready for the challenges that an ever-changing outside environment will demand of you.

Let's go. It's time. Your uterine home is ready to expel you. Note that the pressure about you is rhythmically increasing in frequency and tone and you are being directed toward a well camouflaged opening in the lower segment of your mom's womb. As the pressure rises higher and higher, you are beginning to move downward under the complete control of outside forces.

Slowly, with each uterine contraction, your head is pushing toward the lower segment of this powerful organ, using your head as an effective bougie. The lower uterine segment is beginning to slowly thin and then rise over your head and in this way the door to escape is slowly opening. Escape to where? First into a cylindrical, moist, lubricated chamber called the vagina. It is a rugated organ that firmly massages the descending fetus whose only mature perceptions are smell, touch and taste, probably in that order. The ordeal of passage may take minutes or hours, dependent on the number of children who have passed the same way. When you have traveled about 6 or more inches, the outlet and end of your journey will soon be near.

Your head, which has been in contraction since you began to conform to the upper uterine segment, is now beginning to extend as you descend, as if in anticipation of the new world of dryness, light and noise that you are about to enter. Your head posture has opened your larynx to the air of the world, and you are about to enter a place where self breathing will be demanded of you. The uterus is suddenly not working alone. Mom is beginning to strenuously push and that wonderful lady, who has supported your growth within her, is now ready to bring you into this world and teach you all that is life.

Your head, your shoulders and finally your body and legs have entered this world and with a final scream of delight from your mom, your doctors, and the whole team of helpers who joyfully participated in this magical experience, you are born. As your source of sustenance, the umbilical cord, is cut and the maternal Oxygen is no longer present to serve you, you will suddenly cry and scream and begin breathing, at first haltingly, but very soon rhythmically for the first time on your own. It is your personal way of announcing to the world your birth and the successful first test of self sufficiency.

It is your introduction into what appears to be a world of unfathomable chaos and lies in direct contrast to the peace of several hours before when you were floating in your primal sea, totally dependent on Mom.

Your feelings (inceptions) are totally incomprehensible and have been called infantile chaos (or anxiety). They are the global (coenesthetic) precursors of what one day will become discrete and comprehensible messages from your inner world. They will become your feelings of hunger and thirst, along with your gastrointestinal and genitourinary urgencies. The perceptual receptors, your eyes, ears and skin are yet immature and will undergo a slow myelinization and maturing process over the next two years, so that the first messages your virgin brain will receive are those from our world of within-ness. They are the survival reflexes with which you're born. The veil, or Para veil, is designed for the infant to be able to first differentiate his/her chaotic internal world first and only at a much later date are you forced to see and hear the outside world which is less comprehensible, with all of its intangible complexities.

Yet although incomprehensible, in the short journey from the uterus to the outside world, you have touched, tasted, smelled and been massaged by that rugated cavern through which you've just passed and magically it has left powerful, non-verbal and poorly understood messages of comparison between the idealic world of peace and total dependency and the new world where you've been given the responsibility for breathing; leaving behind peace for chaos; darkness for light; and a warm watery surroundings for a cool, dry atmosphere. On top of that you've been placed in the hands of unknown creatures who will be teaching you all you'll know for the first 12 years of your life.

At no time will your brain lay claim to the virginity it knew before your birth. According to John Locke, human beings are born with a clean slate, a tabula rasa, with our minds being "empty tablets capable of receiving inner and outer imprints but having none prior to birth". The truth of this statement will be argued at another time.

For the next two years and at no time more striking than in the first year, you will be learning and filing non-verbal data regarding your internal feeling world. It will be much later, when your eyes mature, that Mom will introduce the perceptual outer world, introduce language, data, and the meaning and significance of all the intrinsic feelings that caused the infantile chaos in the first place. Their meanings will not only be made clear but also you will learn the importance of your teacher, your Mom, who taught you by taking away the negative feelings of hunger, thirst and wet stinky diapers, and in their place made you comfortable with her embrace, warm breasts filled with milk and wonderful smells, and thus eliminating the first feelings of chaos and changing them into the understanding of an exciting world of adventure and discovery. Although different, it will be nearly as perfect as the tranquil sea from whence you came. Each day designs the next day's competences. Mother will be responding to your every need and want, and continually talk to you so that these needs are eventually given meaning and power and a means of one day communicating intelligently with the overwhelming but responsive outside world that is willing and ready to serve.

The only language with which you were born are your postural changes, your facial emotions and the verbal cries and grunts that add meaning to early language and at times become part of the communication process for the remainder of life. Acceptance and rejection, a reflection of contraction and expansion, are the early means of communicating your feelings.

Despite this preverbal awakening, those internal and external experiences of life are not approachable via psychologic and psychiatric techniques even if your life was pathologically lived, because the route of language had yet to be developed and the experiences, that may have precipitated a wide array of unhealthy feelings, were not verbally attached to cause.

These inner feelings, having been completely differentiated, are now waiting for the time the perceptual apparatus has matured and is ready to introduce the outside world and give articulatable meaning to the infant. As the "Person, Place, or Thing", outside world becomes a part of the child's perceptual awakening, the feelings tones, previously described, are attached to every object so that from the second year on there is a subject-object fusion taking place- 'every object having an emotional positive or negative cathexis'. This underlies many of the problems associated with the embryonal, intellectual unfolding process which requires, in the world of science and logic, a separation of the subject-object fusion. It is historically not until the Renaissance that this separation began to take place in the average adult.

It is for these reasons that the linguistic, mental and rational sciences have failed, too frequently, to solve mental illnesses that grow in number each year and appear to be approachable only by pharmacologically interfering with the neurophysiology of the brain. Although not spoken of frequently, especially in relation to the treatment of psychotic illnesses, the drugs are frequently used to modify the illnesses so that the caretakers can more easily care for these very sick people.

If these and other illnesses are not rationally approachable via the theories of Allopathic Medicine today, what can we do to discover a way of treating those problems in which the early developmental brain has gone astray and becomes the behavioral cause of problems early or later in life?

With this in mind I approached the problem which seemed to be hiding behind closed doors and that in some way had to be opened. Just as the virgin brain of the infant pounded on the uterine door that would open the world to the outside and eventually teach language and the art of Reason, it was my goal to pound my contaminated rational head on the pre-verbal, pre-rational doors of the already closed uterus and contract into my post birth global brain where pre-verbal mysteries lie and discover the passageway back to the beginnings of life where the create/destroy ratio is high enough, and illnesses of the mind or body do not exist.

The pounding process has taken nearly 50 years and the total by-product of this search, although far from complete, has altered my thoughts regarding human behavior, man's innermost feelings, and the Allopathic Medicine I was taught in Medical School."

Jonathan stopped for a while to observe his class. They were wide awake and attentive and he was about to continue when Laura raised her hand. "Jon, I remember in your second book and Treatise, you told a story concerning the Pathway to Love and as I recall, whereas the initial encounter of the boy and the girl was very formal and rational, as their relationship got closer, some of the things that occurred during birth were beginning to take place, only in reverse. It was as if with an increase in closeness, the Pathway of Love is merely a recapitulation of birth only in the opposite direction. Do I make myself clear Jon?"

Jonathan was pleased but not surprised. No-one understood the consequences of the journey from Para dominance to Ortho dominance any better than Laura who was working side by side with Helen for over ten years. "Laura, I want you to read the latter part of that chapter when the boy's control mechanisms are beginning to disappear."

Jon handed the book to Laura who quickly found the place.

"May I kiss you?" he asked, as he impulsively drew her closer to him. For the first time she felt his pounding heart pressed tightly against her. There were no questions regarding her wants, but the conflicts of right and wrong were dissecting her soul. As her mind revolved in agony of search, his heart took up the challenge and her inner feud was won. She grasped his head and drew his lips to hers and then for fear of seeing something less than the passion she felt inside, her eyes closed and their merging needs found outlet in the sucking motions of their lips.

Their bodies cried out 'Touch me, rub me, let me feel your warmth and the wetness of swollen lips'. But this had to stop. The evening ended and the boy and girl, in wondrous discomfort, bid farewell.

There were only eternities, not seconds that clicked the passing moments between their visits. Bound by tradition, bound by the aged codes that warn of sin and pain if the rigidness of right and wrong were not perceived by both, they found the pain of being parted more than they could bear. Thus when equal vows were whispered fearfully beneath the canopy of leaves and flowers, the wrong became rights and the rights became songs. No longer did their bodies cry in need of warmth and tactile pleasure. The sweating smell and touch of nakedness took on the search for needs the lips and tongue had previously not found. Each roundness and elevation, each secret crease and hidden pool of moistened softness; each taut erected hardness of congestion became the hidden prizes for the searching hands that sought them. When beats of heart and pants of breath had reached the point of madness; when touch and taste, smell, sight and voice had all been lost in their diffusion, he placed the hardness of the He within the softness of the She, and merged in pain and pleasure they found the mystery of convulsive labor. Like smiling death, they lay prostrate within the womb, until the cool dry world of reality awakened them to all that is really life".

Laura stopped and the response of the class was memorable. Each one held their own memories of that moment when they first became sexual creatures. Laura was ready to finish. "Note that contrary to the birth story the perceptual journey is reversed. First we see. Then we hear. Then we touch, smell and taste. When there are no resistances, we then merge, contract, close our eyes and in the darkness of our primal world, we rediscover the peace and pleasure of dark and fearless intimacy. It is in the foreplay of intimacy and the magical awakening of the primitive feelings that accompanied the journey out of the womb that we once again discover that which we felt, when we first traveled the postural, pressured, tactile, olfactory and lingual journey out of the womb on the day we were born."

Jonathan was pleased. "I'm glad you remembered that story Laura. It appears that sexuality is the perceptual searching and rediscovering of the pre-verbal period of the infant-maternal dyad. The cataclysmic orgasm is the oscillating, vibratory penile-vaginal storm before we enter into the Para tranquility of the warm, moist primal sea again. The perceptual journey of a fetus being expelled at birth is merely acted out in reverse and the orgasmic moment, which represents the processes of uterine and vaginal contraction when the baby is forced into extension, also forces the sexual adult into an extension that precedes the contraction of the mature bodies as they sense the tranquility first experienced in the primal sea.

It is now important to recognize that spirituality, religion and the concept of God are derived from the pre-rational, primordial feelings of the global infant. The very process of the mother serving the infantile needs, occurs when the infant senses the intense, uncomfortable inner world of the self being served by an unknown outside world that magically responds to its cries, which are the prayers of the frightened, anxious uncomfortable infant. In fact, as we grow, most of our pains, physical, emotional and spiritual have required someone in the outside world to help serve and resolve them.

The circuitry of this Cerebral awareness was designed and templated when feelings were primitive, very powerful and not yet articulatable. It was the time that our passions were templating the subcortical hypothalamus, the hippocampus, the amygdala and the right cerebral hemisphere, the very areas that contain all of the inceptions (feelings) that wreak havoc when we are faced with dangers. They also control the vitality and excitement, when we've fulfilled our goals and responsibilities, or the peace and tranquility when all of our needs have been met and love envelopes us from the maternal bearer of Touch and Embrace. At that moment they are the fulfiller of all of our needs.

All of this happens and grows during the first two years of life and designs those qualities which will be powerful influences on the rest of our life. Who, I ask you, helped create this passionate and non-verbal miracle?

First there was Mom.

Then it was the Physician.

When the sense of mortality enters, the concept of that which would interfere with life, God, was born. But initially, God is Mom."

The class was silent. Suddenly all became clear. The Alternative Healer was merely usurping the role of the Toucher and the compassionate healer, the very role that many Traditional Physicians have thrown to the wind.

When Jonathan quit speaking he expected an immediate reaction from his class, but all he saw were looks of amazement. Finally Kenneth stood up and faced his teacher.

"What you've said amazes me Jon. Do you realize the enormity of this concept and the reaction of the outside world when you make it public?"

Jonathan was very serious when he responded. "Yes I do and this was the reason why I wanted Helen to be sure that what I offered the class, as a powerful postulate, was to stay within this intellectual fortress that we, as a group represent to all of the fields that we now represent. If it frightens you, I understand. As far as you are concerned Kenneth, by all means do not share this with your mom or dad. It would reawaken old fears in Jeremy and he would erupt like Mount Vesuvius.

I'd like to give this time to sink in. The power of Para infancy has been presented to you in a way you've never thought of before. We have one month before we meet again. It is a long enough time for you to incubate this seed that I've planted and I'll be fascinated with what you've all done with it when you return.

I believe it is shopping time girls. Helen, George and I will be checking out of the hotel and going back to Long Island where we have much work to do. Daniel, you continue your work on the design of the Traditional Medical facilities and hold off on the Alternative Systems until we meet again."

Jon was finished and the class had much to absorb because of the apparent attack on tradition and the obvious new powers that the non verbal influences had bestowed on the newborn and his/her mother. The ramifications of this concept touched on nearly every Philosophy, Theology and Human Behavior.

As the class dispersed, Helen noted that Laura and David and Cassy and Steve, quickly paired off as they were leaving the conference room. She turned to Jonathan with a knowing smile and then they both left hand in hand aware of new love stories that were in the making.

Chapter 11

Kenneth and Ally left the conference room hand in hand, yet very much detached from each other in thought. What had happened at the conference reawakened the discussions they had had at least 10 years previously when Jeremy, Kenneth's dad, had requested an in depth discussion of religion. Jeremy, a Fundamentalist Christian, was faced with the frightening awareness that his son was about to marry a Jewish girl and that the family lineage would now be contaminated by someone of a different faith.

This was a decade ago. That long evening of discussion had tightened the bonds between father and son. It made it possible for Nellie and Jeremy to love and worship their two grandchildren when, in time, Ally presented them with Joshua and Esther.

Ally was critically sifting through her mind regarding her maternal life history. She already was very aware of the critical role a mother plays in the unfolding of an infant, but she had no idea it was her responsibility to design the inceptions and the non verbal feelings that would be the seeds from which a mature concept of God would flower. It seemed inconceivable that this seed would continue to grow during each phase of the infant and child's maturation. If Jonathan's postulation were true, to the growing infant God was Mom and every other life supporting savior who rose to the occasion when man's survival, his very mortality, was involved. This was the very template that empowered the Priest, the Rabbi, the Minister, the Doctor and every powerful leader who had been programmed into the brain during the first two years of life.

If ever a question arose regarding the importance and dedication of motherhood, Jon's theory was a condemnation of every woman who delivered a child and then handed over its care and training to a Day Care Center, or for that matter any other person other than themselves. Those women with careers or who chose to focus on an outside job rather than deal with the creative problems of the unfolding infant were avoiding the greatest responsibility a woman could ever undertake. She was therefore sacrificing the infant to the game of chance.

Despite Ally's role as a physician, she did know how very wise she was when she decided to mother her infant and hold in abeyance her active role in clinical Medicine. Both she and Kenneth had felt it best for the babies, never realizing the potential disaster they, by chance, were avoiding.

If the templating of the pre-verbal global inceptions, the within-ness of man yet still not differentiated, was the early responsibility of the Mom during the most Para phase of infant development, what was the responsibility of the father. This was the internal dialogue of Dr. Kenneth.

No one other than Helen and Laura was more acquainted with the Ortho-Para theory. Kenneth quickly realized that if the purist notions of pre-verbal Para were the responsibility of the mother, the father had to be the teacher of the essential nature of Ortho in both the male and female child.

Their trip by cab to the airport and their plane trip home continued in mostly silence as they both contemplated their roles and the significance of Jonathan's revelation. Kenneth asked himself in what way the interaction between the baby and Mom, and baby and Dad, differed? For one, Mom, is relating to an extension of herself, just recently detached, but still emotionally merged in unconditional love. These feelings were practically a physiologic extension of the pregnancy. Unconditional, all encompassing love is pure Para.

The father is not privy to this integrating relationship. He must be introduced, post partum, to a new stranger in his life. If he wants to fall in love with this infant, it is a conditional love that can grow or diminish, dependent on the effect of the infant on the marital relationship.

Whereas Mom is privileged to a pure and untarnished relationship, Dad is as conditional to the infant, as the infant is conditional to Dad.

We can have a relationship in the making, or one that is precipitating evidence of disaster in the future. Dad is not always there. He is not privy to the nude and all encompassing touch and feelings that Mom represents. He rarely represents the response to infant distress except possibly with the changing of diapers. If the baby is not being breast fed, he can be the cuddler and bottle holder during feeding, and if dedicated to the role daily, he may slowly create the unconditionality that Mom represents but it is not automatic.

Dad goes to work. He represents the scary and provocative outside world. He brings home the tensions and anxiety that Mom must cater to and treat, no less than she treats the infant's intrinsic discomforts. Dad represents the tensions and anxieties away from home and as the child grows, the comprehension of this tension becomes a greater reality that he soon realizes he will someday be expected to face. As the child grows, so does the Ortho that becomes part of his life.

Dad represents scare and he has expectations of the child that are far greater than Mom's. Dad gets angrier easier and is not afraid to project this anger on everyone around him.

This internal monologue consumed Kenneth during the cab drive and the one hour flight home, so that when they walked into their home they were both ready to talk intelligently about the subject at hand. When they had made themselves comfortable, Kenneth spoke.

"Darling would you like some wine? I'm going to relax with a glass of Chardonnay."

Allison quickly responded, "Thank you Ken. I sure would like to spend a little time talking to you about what happened to me this morning."

Kenneth smiled "I knew you were in deep thought Ally, just as I was. I'd like to share my thoughts with you also so let me get the wine and you relax. I'll be back in a moment."

They were both anxious to share their thoughts and Ally was the first to speak. "Kenny, although I believe most women recognize their important role in their child's upbringing, I question whether they're fully aware of how vital the role is and will always be. With all of my training and with a deep awareness of not only the work of Jonathan, but also of Mahler, I still was not fully aware of the vital role I played in the mothering of our children and ultimately the life-long effect mothering had on the destiny of all children."

Kenneth quickly responded. "I agree darling. For the first time I understand the unconditional love you gave to Joshua and Esther. I admit I didn't comprehend how you did it when they were naughty and were pushing my anger buttons. As we were coming home the naturalness of both of our reactions became clear to me and it made me proud that both of our instincts, although different, were designed for a budding Mom and Dad. I also know what Jonathan was ultimately expecting from all of his class. It is the Momma and her Para polarity that is the power of the Healer, whether male or female. Those therapies that initiate this Touch and intimate environment have been neglected by the present Medical profession because it's so time consuming, but it is exactly what many of the Alternative Healers are dispensing with compassion and understanding.

With this in mind, within the next month we have been given six Alternative Cares to evaluate, both from their therapeutic technique and the space they require to practice their art. Do you have the list darling?"

Allison was pleased with this rapid summation of what they had learned about themselves. "I'm beginning to think that the time those Healers give; the understanding and compassion they show, and the depth of conversation regarding their patients' system conflicts they listen to, plays the most significant role in altering the Autonomic chemistry and in that way changing cellular breakdown to creative buildup.

We've been given Reflexology, Massage Therapy, Aroma therapy, Herbal therapy, Acupuncture, and Moxibustion therapy. I must admit the goal we'll be seeking is first the eradication of our abysmal ignorance regarding all of them. Of the more than one hundred therapies on the list, at least the one's we've been given touch on Medicine and may, via Jonathan's explanations, become the main examples as to the value of this approach to illness. But we must also remember that Physicians were not the only ones that were sought for the treatment of the ills in past ages, so that we mustn't consider Alternative Medicine as a brand new discovery as any physician of Chinese Medicine would surely tell us."

Kenneth was quick to nod his approval.

"You know Ally, I keep reflecting on Jon's description of the birth of a baby and its relationship to sexuality. But far more significant I feel is how it affects our knowledge of the many therapies we're evaluating. Consider the sequence again. With the onset of labor there is an increase in intrauterine pressure. It is the first sign that the pregnancy is ready to terminate. That awakens in my mind the therapeutic effect of Bundling. Remember, we found that shortly after Joshua's birth, whenever he became fidgety, even though we felt all of his needs had been met, if we wrapped him tightly in a blanket and squeezed gently he rapidly quieted down and went to sleep.

Then there is the downward pressure of the head as it acts as a bougie, dilating the lower segment of the uterus. Cranio-sacral therapy applies pressure to the head and neck and the Chiropractic practitioner performs the same maneuvers on the spine and pelvis, relieving headaches, neck pains, back pain and the radicular pains to the lower extremeties.

As the infant moves down the birth canal it is constantly being massaged by the uterine muscles and the rugated mucosa of the vagina. This massage leaves subliminal messages that become the underlying reason why massages of all sorts feel wonderful and relaxing.

As the baby descends its head is extended and the primitive olfactory system is exposed to the secretions of a well lubricated maternal organ. Since the olfactory system is one of the first perceptions that mature, even before delivery, we shouldn't be surprised that certain smells have very positive effects on our behavior. You only have to consider the pheromones that play so powerful a role in generating aggressive reactions in the sexual encounters.

The most sensitive, sensuous areas of the body are those where muco-cutaneous junctions exist. It is where the outer skin is in contact with the inner lining of a body orifice. We're talking about the lips, mouth, vagina, anus and even the ears, all of which are the holy, holy parts involved with love and intimacy. Tell me what is more therapeutic for all illnesses than love?

Pressure, touch, massage are all born when the infant is in pure Para and yet un-awakened to an awareness of the outside world and therefore un-encumbered by an awakened Ortho.

When I realize how significant is the birth process to all creatures, and how it ultimately influences our lives as we move from that pure Para environment of the amnionic sea into the world of Ortho and all of its challenges, it's hard to believe that it eluded the pundits of Medicine for 5000 years. I realize that some of the oriental ancients touched powerfully on their therapeutic value, but did not understand why it worked."

Ally was fascinated on hearing once again the relationship between the birth story and the various approaches to healing.

"Kenneth, you, Cassy and I, along with Jonathan and Helen, will be more influenced by this intellectual revelation and revolution than anyone else in our class. I will try to set up evening appointments for both of us with the well known therapists in the specific fields allocated to us and hopefully they will not interfere with our daytime responsibilities."

Chapter 12

When Becky and Dan left the meeting, they went straight to the car and headed out to Dan's home that was close to his new architectural project in Long Island.

Their ride home was unusually quiet, as they mulled over the significant postulates that Jonathan had shared with them. Although the meeting had been called to benefit Dan and to give him essential structural information for his Ortho-Para Research Center, Becky was somewhat frightened over the significant way this postulate might affect her role as the Minister of her church.

If what Jonathan said was true, that the birth of the concept of God occurred during the pre-verbal interactions between mother and child, would this postulate have an effect on the parishioners of her church? For that matter we were talking about all religions and all ministers of God. It once again sounded like many of the philosophers who claimed that God did not create man, it was man who created God. This notion had always haunted Becky and it became a bigger problem when she was advised by an agnostic friend to read a book by Matthew Alpert entitled "The God Part of the Brain". After an exhausting review of most of the major philosophers, Alpert concluded that God, as most of early organ conformations, could be attributed to a Gene- the God Gene.

Like many other probing minds throughout the Millennia, Alpert asked the question "Is there a God? If he exists, is he in some celestial residence that we have to wait to see when we die? If he truly exists and accomplished the miracles of creation, why would he have done such a poor job with human behavior so that in His name, people have been killing each other since the beginning of human time?

These were good questions and the answers have been many and as diverse as the number of brains that sought their answers. In his book, Alpert takes us on a journey through history and touches on many of the great Theologians, Religionists, Philosophers, both Eastern and Western and he logically comes up with an amazing conclusion. God did not make the world and mankind. It was Man who created God.

He then offered the postulate that in time God became a genetically controlled piece of Man's being, with structure and wiring that evolved because it served a powerful function.

One day Man reached that point of maturation which allowed him to become aware of himself. Self awareness precipitated a concern that at some point in time he would be no more. From this point on, primitive man needed to evolve a circuitry that would ultimately help him cope with his mortality and to accept an eternal life, if offered it, with the condition that he follow the precepts laid down by those most sensitive clergy who devoted their life to the service of the Lord and knew the road to eternal salvation.

Having come to this conclusion, and having made God the product of an evolutionary event, he then chose to place the mechanisms that made Spirituality and God, so Universal a Truth, a part of the genetic code. Having accomplished this, with a very erudite and scientific process, he successfully took God out of Heaven and put Him into Man's brain.

It was once again pitting the battle between the Preformationists, who believed that man was born with everything built in, against the Epigeneticists who believed that although man had built in qualities, his environmental experiences altered these qualities and the by-product of the interaction between that inherited and that created by systemic interactions, ultimately became the dominant influences on his life and behavior. To Matthew, the genes had become the All. The power of Parents, Work, Goals, Morality, Ethics and Creative thinking had finally been displaced by genes. No longer were the powerful forces that helped mold man considered relevant because they were substituted with long chains of nucleitides and their Purine and Pyrimidine bases. For the sake of placing God where he belonged, the power of genetics had replaced the power of religion.

When Becky read these conclusions she had immediate insight into Matthew's problems. Like all scientists, he felt that everything could be explained by science. The power of the intuitive, the passions and the spiritual that were incomprehensible via reason, had to ultimately be abolished. She could easily brush these journeys of Alpert's logic out of her mind because she knew, deep down inside of her, that it was not genes that made her turn to God and her faith.

Now a serious dilemma was presented to her by the man who she believed in more than any other person on Earth. Jonathan's claim that, Spirituality and God were derived from the pre-rational feelings of the Global child, were more difficult to dismiss. It was Jon who explained that the pains of hunger, thirst, stool and urine control, that originate within every infant, have always required the outside world to resolve these intrinsic problems. It was the same with most of our pains, physical, emotional and spiritual. They always required someone in the outside world to help resolve them. So now it was with one's mortality. The fear that comes from within was best resolved by an outside force that could simply explain it away and introduce eternal life as a possibility.

Why and How?

Jonathan had presented the thesis that it was during the chaos of the pre-verbal period of infancy that an unknown, omniscient, omnipotent figure arose, who assuaged our needs, protected and embraced us, loved and explored our bodies and tranquilly won over all of our hungers and thirsts. It was this image that became the all powerful loving figure that remained part of our existence. It was not, as Matthew postulated, a gene that had been implanted in our genome. It was merely Mom, who played this role only during the first 12 years of life. The infant was truly born untainted by any concept and it was its surrounding environment and the systems of interaction that surround him from day one, that programmed the untainted brain with its first awareness of the world of Mom within which He would grow.

Why don't we eventually know this? This God figure is created before we can speak or reason. It is a pre-verbal awareness, a reflection of that which is sensed even before language is comprehended. We sense that we will be always taken care of by a powerful, spiritual representation of the outside world that must always be there to calm our fears, obtund our pains, embrace and subdue our anxieties, service our survival needs and tolerate our foolish and dangerous experiments, knowing always that the one who oversaw our creation would always be there to successfully counteract our destruction.

The circuitry of this sub-cerebral awareness took place when feelings were primitive but very powerful and not yet articulateable. It was the time that our passions were templating the Right brain and the sub-cortical Hypothalamus, Hippocampus and Amygdala, the very areas that contain all the inceptions (feelings) that wreak havoc when we are faced with danger, and peace, pleasure and tranquility, when all of our needs have been met and love envelopes us from the bearer of Touch and Embrace, Mom, the fulfiller of all of our needs. All of this happens and grows during the first two years of life and designs the qualities that we have eventually given our God. God made it possible to know him by creating a love for Mom that could eventually be transferred to Him when the child matured and became aware of the magnificence of the Universe and of Life.

As Becky pursued these thoughts, she immediately recognized the difference between a genetically implanted God, and a God that is discovered at the same time that Love is found.

Daniel was very sensitive to the change in Becky's behavior following the lecture and he quickly understood the imposed professional pressure that Jonathan's assertions represented as far as religion was concerned. In no way had Jonathan ever demeaned the many religions of the world, but this was certainly not the case in the writings of some of the major Philosophers of the world.

The battle for the minds of men was the turf on which Philosophy, the Theologies, the Bio-physiologists and the Psychologists played their abstract games, woven of words, thoughts and occasional eurekas that became the new paradigms designed to make the world's dramas more comprehensible to the students of life.

It appeared to Daniel that the Ministries of all faiths, as interpreted by Jonathan and understood by Becky, were to become just another form of Alternative Therapy. Dan invaded the prolonged period of silence. "Becky, I can see that Jon's pursuit of Para has invaded your turf and you're baffled at the turn of events. Am I reading you wrong?" His concern and the airing of her internal stress brought tears to her eyes.

"I'm confused Dan. Is he saying that God is Mom and there is no other God other than the deposited global messages of infancy? Is he saying that God has become the non verbal passion of a powerful surrogate mother, when Mom eventually becomes a balanced reality and a new omnipotent and omniscient image is needed to face the uncertainties of the future? Am I merely the conduit of the priestly fabrications of past millennia, men who found a unique way of gaining power, personal security and the means of taming a feral people who socially remained in Jon's Endophilic Era of Civilization?"

To Dan, this was an amazing and unfortunate turn of events that was totally unexpected. Most of the questions were logical sequelae of Jon's postulate, and Dan, who was so close to Jon, both in faith and in their ultimate goals for the Ortho-Para Research Center, felt it was his responsibility to respond to Becky's crying pleas.

"Becky, I am as certain about what I'm going to say as I am certain about my love for you. Jon is not an Atheist or Agnostic. He is a God fearing Jew, who has struggled to understand the origin and meaning of God. He has come to the conclusion that since the infant, at birth, is born empty of any concepts and that knowledge cannot be wired into the infant's brain, he wondered 'How does it get there'? There had to be a Universal process of programming the awareness of God into every human, of every culture in the entire world. It had to be non verbal, because God, to all of us, has never been verbalizable, despite the enormity of this passionate, non verbal pressure. What more powerful way could He implant the concept of God than through the non verbal period of contact with Mom who represented the purist love that anyone would ever experience. Becky, God is love because Mom taught it to us when we just entered the world and had to be dependent on a God-like creature who would forever be God's ambassador to his children. I believe that the Ministries of all faiths were created to oversee each individual of faith and guide him through moments of despair and disappointment when God seems to be unreachable because of the intensity of Man's feelings of isolation."

Dan had searched into the deepest depots of his heart and mind and succeeded in touching the heart of Becky. As she listened to his explanation it became clear that her role with her congregants was to lead them through the miseries and disappointments in life, no different than Mom did when she was the prime and most important teacher that any child would ever have. Mom turns out to be the teacher, the nurturer, the caretaker and the healer, all wrapped up into one package.

By the time Dan had vented his feelings and his interpretation of Jonathan's theory, Becky had recovered from her acute, momentary depression. She approached Dan with endearing eyes. "Thank you Dan for being a little more left brained than I and explaining what, I'm sure, is an accurate interpretation of Jonathan's views. You've placed me back in the position I knew I represented to my flock, especially the young ones. Have you thought at all about the clinical research we're responsible for in the next month? Did you write down the therapies we are supposed to evaluate?"

Dan was overjoyed with the rapid disappearance of Becky's fears. He responded quickly.

"Yes Becky, I did." He began searching through his pockets and found the note in the breast pocket of his shirt. He slowly tried to read his own handwriting and finally with a smile of success, responded to her question.

"Becky, we have the following therapies to evaluate, Spiritual Therapy, Charismatics, Divine Therapy, Flower Therapy, Love Therapy and Self Healing Therapy."

Chapter 13

When George and Cassy left for the meeting in New York, in response to Becky's clarion call, they decided to take his car and take a one week vacation from the University in which they both had professorships. George was teaching Oceanography, a course which had never before been offered by the University. Cassy was the newly groomed Professor of Cardiology. When Kenneth and Allison both decided to go into General Practice, Kenneth with emphasis on Ortho-Para Medicine and Allison with a focus on Gynecology and Pediatrics, Cassy felt that she had the responsibility of representing the Specialties, and she chose Cardiology because it represented the specialty devoted to Ortho diseases, while Allison was to be the representative of Para diseases.

It was all part of the initial plan that was agreed on in High School and this focus and intent was carried into each one of the 17's individual professional choices. Following the conference, they left New York, via the northern route, intent on driving across New York State on the Freeway and seeing the Northern Catskills and Southern Adirondacks on both sides of them. They looked forward to seeing the huge wine country they had heard so much about and some of the areas that George remembered as a child when he went to the Catskills by choo-choo train to visit his grandpa who vacationed each year in the small town of Parksville. His grandparental memories were pure joy, more like a fantasy that could never have happened, but they did. George had gone into Oceanography because his dad had been a sporting Scuba diver and seemed to go into reverie whenever he spoke of the watery depths. As a high school student his dad took him on a deep sea diving expedition and taught him the basal pulmonary physics essential to the knowledge of all lovers of the Benthos- the home of all bottom dwelling sea creatures.

When Cassy and George discovered each other again, after they had established themselves in their respective fields, they had both accepted teaching positions at the same University and at the time of Becky's call, they were leading professors in their respective fields.

Initially, when they began their trip home from New York, they were quiet and chose to watch the city scenery and traffic as they wound their way to the Hudson River Drive which was to become their northern route until they turned west onto the New York State Freeway.

Cassy broke the silence, "George, what did you think of Jonathan's last lecture?" She had a deep questioning look on her face.

George was slow in answering and appeared in deep thought. "That's what kept me quiet for so long Cass. I kept watching Becky as she heard words that seemed threatening to her basal beliefs and of all other religions, and it was especially most evident when he ended one sentence with 'Why of Course. God is Mom'."

Cassy quickly replied "I watched Becky's face also and you're right. I think she felt attacked and was hurt and confused. But I know Jon, and although I might be wrong, he was not negating the existence of God but merely teaching us why our awareness of him is so powerful despite his eternal invisibility." Cassy had gone through too much with Ken and Ally, and was very aware of Jon's and Helen's faith problems when they were struggling with their decision regarding the religious education for their children in their mixed marriage."

George understood and after reflecting for a moment on Cassy's conclusion chose to change the subject. "You know Cassy, I'm so distant from Alternative Care on a day to day basis, especially while lecturing at the University. But when I go on my yearly summer water safari back to the sea, I'm very aware of the illnesses that are found only in relation to the sea, and to those appliances that have been added to some of the major medical facilities in our country, when they have patients who have not followed the rules and regulations that must be followed by every one who chooses to share the beauty and thrills of the sea.

When Ally was discussing the evolution of layered Man and the physics and physiology associated with the elements and molecules, the first thing that entered my mind were the many illnesses associated with the changes of pressure when we dive in the sea or when we fly high in the heaven. What I'm saying is that despite the fact we've been given Alternative Cares that we know so little about, I'm going to convince Jonathan and Dan that a Research Medical facility cannot neglect the problems of the water depths. Since the Institute is right on Long Island Sound, it's in a chosen spot for this type of study."

Cassy interrupted him, "But George, the illnesses and therapies you're talking about are recognized by Modern Western Medicine and have been used by my colleagues in Cardiology for many years in their Oxygen futuristic studies, especially in poorly responsive Angina."

George replied, "I'm aware of that Cassy, but the relationship between blood pressure; intravenous Gas pressure, temperature and volume; pulmonary gas pressure, temperature and volume; and the same physical parameters in the kidney, liver and heart vascular systems, and their relationship to an Ortho/Para ratio, has never been done for the various atmospheric pressures we face when we're exploring the depths of the ocean, or reaching for the heavenly bodies in our solar system. In a sense the Ortho challenges of Physical, Emotional and Intellectual research should be part of our goal so that we successfully reach out to new frontiers, both in the sea and the sky, so that the dangers to our exploring frontiersmen get less and less."

Cassy started giggling as she cuddled up a little closer. "You know darling, your Ortho dominance reaches out for answers to questions that only Science and Reason can solve. I sense your Para potential rising only when you jump into bed with me and throw that Ortho, problem solving mind away. I'm sure when you're diving amongst the beautiful creatures of the deep, they must precipitate a Para awakening in you, a passion for their beauty, and in what way they fill the visual sensuous needs we all have. But George, you are my Autonomic balance and I'm glad you felt the same way when you chose me as your wife 5 years ago."

By the time this conversation ended they were driving through the wine country of New York State. On both sides, as far as they could see, vineyards were in blossom and their beauty brought an end to their talking about anything other than the landscape, with the mountains to the south, bringing wonderful memories of George's childhood in the Catskills, and to the north, the higher Adirondack's where he took Cassy on their honeymoon on the shores of Mirror Lake in Lake Placid.

When they had passed the beautiful scenery, and just grassland passed before them, Cassy began pursuing their goal again. "George, what Alternative healers were we given for our month of homework?" She smiled as she thought that Jonathan continued to find challenges for his class for 25 years now, and the challenges were never easy but always interesting.

George revealed a piece of paper and handed it to Cassy to read.

"Well it appears that we have the following six . Allopathic Medicine and its specialties, Ayurvedic Medicine, Biomedicine, Chelation Therapy, Curanderosa Medicine, Environmental Medicine, and Enzyme therapy. I think that your comments also added high and low Pressure Medicine.

That's a big job George and I see our evenings monopolized by our dear teacher. It is truly an important undertaking and the meeting next month should be very interesting.

George agreed and the rest of the journey back to Ohio was scenic and uneventful.

Chapter 14

It was nearly four years since the stunning red hair of Laura was seen beneath the canopy of flowers, the ancient Chupa, holding hands with the classes chosen spokesman David. It had been a magnificent day of joy for the entire class and just as Helen had predicted years before, there was a pairing off taking place that would eventually see eight couples, bound by the mutual dedication they made when they were just high school students.

When Laura had completed her PhD in psychology she had already been working with Professor Helen Fentonowsky for more than ten years and had reached the frontier of Ortho-Para Research. She was seeing patients and was working on the tedious task of finalizing the Ortho-Para test (OPT) and accompanying Script. This would one day be published as the Clinical catalyst for the ultimate dissemination of Ortho-Para Psychology to other major Universities.

Although she still lived in the shadow of her mentors, Jonathan and Helen, she was the primary inheritor of this new Psychologic challenge to the understanding of human behavior and its inner control mechanisms.

In contrast to Laura, David's studies carried him to the frontiers of the intracellular and intra-nuclear forces that powered the unfolding of human structure and behavior. The fascinating way in which the outside environment touched on the nucleic acids that eventually created DNA was an ever evolving confrontation that Laura and David had brought into their professional and marital lives.

Laura had become more subdued than she was during her professional schooling. The sad crises of inter-relational conflicts that she was exposed to everyday, slowly took its toll on the bubbly child that had captured Jonathan, and eventually seduced David into what appeared to be a life-long relationship.

Shortly after leaving the hotel, David was driving toward Great Neck Long

Island where they lived and worked. David turned to Laura, "Babe, what did you think of Jonathan's insights into the therapeutic processes at work with the various modalities now called Alternative Care?"

Laura smiled as she answered. "I guess I should apologize to you, but Helen had already shared it with me about a month ago and I promised not to say anything until today. What do I think? I think the power of the non-verbal first two years of life has finally been given its due.

Words to philosophers are like numbers to scientists. Silence and quiet introspection has never been given the recognition it deserves. When the concepts of absolutes were challenged by the theories of Einstein, the unknown and mystery was suddenly recognized as a significant partner of knowledge. Helen and I were always aware of the power of the inarticulate when we discussed the Para passions that men and women wrote poetry about, but frequently fell short of describing the inner feelings that good poetry must capture. When Elizabeth Browning wrote 'How Much Do I Love You' she succeeded in waving a net of beauty about Love, when just the word 'Love" fails to project the same meaning. So, in answer to your question, Jonathan remains the genius who guides us forward in the intellectual journey that Ortho-Para Research has laid out for all of us." Laura spoke with no evidence of questioning. In her own mind she knew that before Jonathan presented new postulates, they had been incubated and put through his and Helen's cerebral sieves and what came out had a purity of thought that she no longer questioned.

This however was not the mind set of David who felt he was also working in a field that was possibly a Preformationist's dream. The Epigeneticists believed that Genetics played an essential initial part of initiating inborn and reflex behavior. But as Laura would point out in her arguments, it was the environmental pressures associated with living that finalized the process and played a huge role in creating the human differential.

David didn't openly argue that point, but deep inside he was looking for that something the Preformationists still felt existed. It was they who claimed that Man, and all of his skills, intellect and potential for the future, was built in, and that Man was destined to walk a path that he was unable to diverge from. It infuriated Laura and only brought smiles to Helen and Jonathan who encouraged David to dig as deep as he could to prove or disprove these contentions. At 35, that was what David was doing and his beliefs still hung in balance.

David replied "Laura, I love you and know that you make my life more interesting and complete. Having said that, I don't think even Jonathan believes his postulates should be taken as a truth until they have been rigorously put to realistic tests. Remember how skeptically we approached the theory regarding the Universe's beginning? I am approaching the concept of Preformationism with a high degree of skepticism, but look carefully at the human and other creatures and their layering to see how powerfully genes can be in control over all unfolding processes associated with life and pre-life. I want to assure you, once again, that my views are not fixed dogma from which I dare not stray. For that matter I was surprised, maybe shocked, by Jon's concluding Statement 'God is Mom'".

Laura immediately began laughing. "I can imagine the furor and possibly terror the Patriarchs of all major orthodoxys would reveal, if they had heard that acclamation by Jon. But you and I know it was his way of shocking us. What better way could the good Lord have of first differentiating the inner incepts, uncontaminated by words or perceptions, than by creating the Para veil. It was only after all the feelings had been properly placed in the non-verbal, subcortical layers of the brain that the Para veil would be lifted and the process of perceptual differentiation would then take place although burdened with subjective contamination during the initial process. It was only after the object-subject separation took place that the deductive processes of science have had any chance of making headway."

David smiled as he answered Laura. "You know that everything Jon, you and I have said, since we began this conversation, relates to how the environment teaches the non verbal infant and that this teaching appears essential to the programming of the brain. Furthermore, our clinical Psychiatrists have said that if you don't expose the perceptual senses to the potential, outside perceptual stimuli, during a very specific and narrow period of time that is bounded for every sense, that the stimulus to maturation of that perception is no longer effective and the perception, so environmentally insulated, never matures.

Assuming this is true, and it has been confirmed over and over again by developmental physicians, why do you worry about my focusing in areas that are pre-rational, pre-intuitive and yes even pre-cephalic, as I continue to try to understand the germ cell's brilliance that ultimately unfolded into a beautiful, brilliant, charismatic red head like you?"

By that time they were both laughing and their journey home was close to ending.

"Yes David, you're right. I shouldn't worry, but somehow I can't help it. I keep thinking "what if Dave finds we are all pre-determined? What if the evidence suggesting we have the freedom to make anything we want of our life is wrong? What if every drunk, dope addict, vicious criminal and thief were all due to a genetic crap game, and Love, responsibility, parenting and a belief in peace, honesty and morality was a lot of hokum designed by money grubbing teachers, priests, ministers, rabbis and the medical profession whose very existence has, in the last 25 years, encouraged money grubbing lawyers and politicians to preach laws and then rob us whenever we succumb to our own pre-destined greed, lust and dishonesty." As Laura said these things and saw David's look of disgust, she suddenly realized a truth that Jon and Helen had taught them long before they had reached the Age of Wisdom. She turned to

Dave and her tone had quickly changed.

"Dave, why am I haranguing you? You're absolutely right. Every babe that is born is a feral savage who must be trained in the social morays of their culture. If this does not occur, all of the sins of man are built into the egocentric, demanding and dependent infant and that is a genetic reality that I've always known about but never carried it down to the germ cell".

They had reached home and quickly got into their respective routines and sat down together only when supper was ready. Laura spoke first. "Dave, did you copy down the Alternative Healers that were given to us?" Her voice did not emit any enthusiasm.

David laughed and dug into his jacket pocket and pulled out a list. It read- for David- Vibration therapy, Magnetic therapy, and Applied kinesthesiology. For Laura, Chinese Medicine, Orthomolecular Medicine, Environmental Medicine."

David continued. "Well, it's obvious, even before I do any of these healers, I'm going to have to find out what and how they do what they do. You know Laura, this may turn out to be a very interesting month."

Laura looked again at the list. "I'm glad I got Chinese Medicine. When I talked to Shang at our last meeting he talked about a wonderful meeting that he attended when his dad, Si Shu, had a long session with Jonathan, comparing Ortho and Para to Yang and Yin and found a rather miraculous similarity between Yang and Ortho, and Yin and Para."

The telephone rang and Laura got lost in a long conversation with Cassy who was struggling with a patient with a difficult Para depression. David went into his study and remained there until bedtime.

Chapter 15

It was still early when Dennis and Cynthia left the hotel. Their flight back to Ohio was not until 7 A.M. Dennis was now in charge of the Centerville Astrophysics Observatory located on the top of Alabaster peak, that overlooked the valley that had been carved by the rapid flowing Ohio river. His life, when at work, was a very isolated life, with a reversal of his nyctohemeral rhythm since most of his studies were spent either observing the heaven for new galaxies that were ripe for discovery, or in contrast, working as a colleague in the University's new Particle accelerator building, that vied with Switzerland for dominance in microcosmic research.

It was in the microcosmic study of subnuclear particles that Cynthia, who had her PhD in Accelerator Design Physics, would eventually meet Dennis again and develop both a romantic and professional relationship with one of the beauties of his grade school class. It was five years prior to this meeting called by Becky, that Dennis and Cynthia were married and celebrated by the entire class. It was a marriage of the Macro-cosmos to the Micro-cosmos. It took place in the large reception parlor of the Astral Observatory, with walls that had been decorated with an inorganic panorama beginning with an invisible haze of mystery, the coalescence of Mattergy, the slow development of a ball of undifferentiated energy, the molecular chaos of a monogalactic ball, the Big Bang, followed by the creation of a multi-galactic heaven and then the segmental development of an expanding multi-galactic system of stars. Within these stars were planetary systems which had the same particle progression from sub-quarkian particles to eventually the electron and protons that were the play-toy of Cynthia.

The Micro-cosmos was marrying the Macro-cosmos and in their joint studies they were finding, at age 35, that the theory of Cosmic evolvement that played so creative a role in their lives when they were only ten, was proving to be closer to the truth than they ever imagined. Of the two, Cynthia was the more verbal. When outside of the Johnsonian Center she asked Dennis, "Would you like to visit the New York Planetarium?"

Dennis was quick to respond. "No Cynthia, I would like to drive out to Long Island and visit the site where the new Research Center will be built. Dan has a really challenging job ahead of him and although he has to design space for a hell of a lot of Therapeutic Healers, I think his biggest job will be to use the land to create a unique bit of architectural magic. I want him to make us all proud. I spoke to Dan just before we left and asked his permission to visit the site. He was overjoyed and will meet us there with Becky. Becky is flying back to Chicago at about the same time we leave for Ohio so that there shouldn't be any difficulty getting to LaGuardia on time."

Cynthia was pleased. "Do you know how to get there Denny? I understand the traffic can be horrific." There was a bit of concern in her voice.

"Don't worry darling. This is Saturday so that the daily traffic congestion is not as great. We'll take a taxi and get a pre-arranged price so that we won't have to worry about the driver taking a circuitous route to our destination."

Cynthia laughed and agreed and within a few minutes Dennis got a driver who was willing to drive to the site, wait for a while and then take them to LaGuardia for one fee. They were soon off on their trip through New York traffic and out of the city.

Dan was waiting at the site when Denny and Cynthia drove up. He was standing on the rocky shore just overlooking the Long Island Sound. The waters were clear and tranquil on this sunny day. The acreage, adjacent to the Sound was undeveloped, but it was a landscaper's paradise with its various wild trees and flowers throughout large groves and patchy grasslands. It was a color treat. Trees, flowers, grass and rocks, bordered with the aquamarine water that slowly became a deep Prussian blue as the water deepened and reflections of the sky and the watery benthos merged for the pleasure of the human eye.

As Dennis and Cynthia got out of the cab they were greeted by Dan and Becky. Dennis, looking with some surprise at the surroundings, spoke. "My God Dan, did you find this spot?"

Dan smiled "No Denny, Jon and Helen spent many months looking before they found this acreage that will be easily reached only when some new roads are constructed during the design process. In truth I have the outer architectural design of this multi-systemic project complete in my head, but I'm at a loss regarding the Alternative Cares that Jonathan will want included. His Ortho-Para Research Center has evolved into a center for the integration of all medical philosophies. As his theories of Ortho-Para medicine evolved and his anger over the slow and insidious loss of Allopathic Touch Medicine increased, the Alternative Cares became more significant to both Jonathan and

Helen. I'm sincerely hoping our class sets me on the right track."

Dennis sensed a little anxiety in Dan's voice and responded. "Dan, if you already have designed the outside in relation to this beautiful property, you can't go wrong. Assuming that your financing will take care of motor driven walls, you can create any variation on the theme of rooms, including one huge conference theatre created by the integration of many small rooms that either sink or raise their walls."

Dan smiled and responded, "I thought of that Dennis, but we're talking about over one hundred of these Alternative Care healers and they seem to grow in numbers on a monthly basis."

Dennis laughed. "Hasn't Jonathan told you why?" Dennis asked with a look of surprise.

Dan embarrassingly answered "I haven't even shared with Jonathan my worries. Do you think I .....should? You ask that question as if the answer is obvious."

"It is Danny. It is. Remember the laws of Systems and Particles we learned in fourth grade. If I remember it correctly law # 5 states that every system, in the process of expansion, contains birth particles in the state of contraction. To remember this always, look at the expanding Universe and the stars that are birthing themselves in all of the galaxies we can observe. For that matter, as the medical system has expanded only God knows how many specialties and subspecialties have contracted into existence and now the same thing is happening in the Alternative Care field."

Dan had a joyous look on his face. "Why Dennis of course, and if the medical system continues to be subsidized, it will continue expanding in all directions and my real problem of design has to be a dynamic one capable of reacting both to expansion and contraction. Denny, the major answer to my problem is solved and the way I handle it will hopefully be adopted by other hospital architects who every decade face a different problem relative to this expansion and contraction".

Becky was just gleaming as Dennis was talking, but she saw something that bothered her in the movements of Dennis' feet and hands. They were in perpetual, meaningless motion. Dennis' speech was fast and he had some stuttering, which seemed to have become a constant part of his speech delivery. What was equally strange was in Cynthia's behavioral patterns. She seemed to be hyper-energized by Dennis' activity, so that in their presence she and Dan were becoming uncomfortable. It frightened Becky and with these observations she suddenly realized that her two dear friends had become what Jonathan had described as Ortho extremes. The end result of this functional dominance, both in males and females, could be early Coronaries, so that Becky knew she had to talk to Jonathan about her fears.

Time flew by very quickly and it was time for them to get back in the taxi and take off for LaGuardia.

"Dan, we have to get to the airport at least one hour before take-off, so we'll have to leave now. It was wonderful seeing you and Becky and I know we'll be seeing you again in one month. I hope our little talk helps you in your project."

Dan was overjoyed and hugged Dennis for his gifted solution. "Thanks Dennis for seeing the solution so quickly. Cynthia, give him some tranquilizers. He moves around like a hyperactive child."

Cynthia laughed but with some reserve. She saw Becky's interest in Dennis' movements and realized, for the first time, that it wasn't normal.

They entered the cab which headed straight for LaGuardia which was only 20 minutes away.

It was when they were seated in the plane and buckled in, ready for take off, that Cynthia asked Dennis for a list of Alternative Cares they had been given. For reasons that she didn't understand he had become suddenly glum and handed her the list without saying a word.

It said, Dennis- Meditation, Neural Therapy, Aerobic Therapy. Cynthia- Cold Laser Therapy, Homeopathy, Pharmaceuticals.

Chapter 16

Steve and Ariana were amongst the last to leave the conference. Their relationship had been consummated only one year previously and they were still in the throws of facing the realities, and destroying the fantasies that all newly weds carry into their relationship. It is understandable why when one considers the very different fields they entered.

Steve was a Geologist who looked at the Earth as a living structure, unpredictable, one moment peaceful and tranquil, allowing life to form and pasteur in its valleys and on its mountains, and painting the landscapes with fertile forests and fields of edible and beautiful foliage. On the firmer surfaces near water, huge cities could be built, for this was the visible and cosmetic outside of the Earth. But as Steve frequently explained, the inner Earth, the unpredictable side of our planet in this and every other solar system, there was an inner turmoil that was unpredictable and diastrophic in its moments of fury.

Ariana was now a Constitutional lawyer. Her concepts were not of the Earth and its unpredictability, but of Man and Woman and their behavior as they moved through the cycles of growth within a social edifice. Her bible, even before she entered law school was "The Federalist Papers" containing the brilliant writings of the founders of our country. She was aware of one commonality between her field and Steve's. The Federalist's arguments were conceived to define the social cosmetics that were essential to a diverse society in which peaceful coexistence was essential to the survival of us all.

But beneath the social outside were the raging egocentric drives which are seen in every infant, and which, if not tamed, can cause as much local damage to an individual and family as the violent earthquakes can produce in a village or city.

To Steve, the Earth was living. It was his life's purpose to find out how it could be made more compatible with all of life on Earth and to discover the essential balance between a Living dynamic Earth and one that accepts and loves those creatures that were born of and on its soil. In a sense the concept of Layered Man was very much a part of his total world concept.

Ariana, in her unfolding, became more and more verbal, and in personality very much like Laura. She was entranced with the translation of the Para and Ortho concepts into one that related Para to Loving and Ortho to Living and that Man and Woman had to create a Living-Loving balance to remain in personal and social equilibrium.

Steve, holding hands with Ariana as they left the hotel, asked "Do you want to take the train to Baltimore or fly? It makes no difference to me dear. It's been fun to break away from my work, but I'm ready to begin again."

Ariana had already made up her mind. She had a class to teach on Constitutional Law and wanted to prepare for it. This year she was teaching a marvelous group of graduate students who took great pleasure in probing the deepest recesses of her mind and she enjoyed the challenge. The subject that would be discussed was Individual vs National Rights.

This was the first class that had challenged her with this question and she knew it would take her some time to prepare for it.

"Steve, let's take the plane. I'll have to review some of the Constitution before I undertake my writing of this Monday's lecture. As I recall there have been some changes in the Supreme Court's interpretation of Individual vs National Rights and it may, in some ways, be relevant to Jonathan's Research project."

Steve was very agreeable "OK dear. There is a plane every 1/2 hour, and if we get to the airport in one hour, I'm sure we won't have to wait long."

They took a cab directly to LaGuardia and within 1/2hour they were on their way home, and within one and 1/2 hours they were sitting in their condo overlooking the Homewood Campus at Hopkins.

Ariana had majored in Constitutional law after taking a post-graduate course on the Federalist Papers. She was completely enamored with Hamilton, Jay and Madison and knew that they represented the men who successfully achieved their goal of creating a Constitutional Republic that incorporated all of the original 13 states. They had made their place in history and had successfully created a Nation capable of defending itself against all adversaries that were existent at that time. Little could they have realized the eventual immensity of a United States that incorporated all the lands from sea to sea, between the Atlantic and Pacific.

The miracle achieved was expressed again and again as they analyzed the history of a millenium of experiments in Democracy, analyzed their potential flaws and reasoned that the solution to the flaws was the creation of a Constitutional Republic which corrected all the potential flaws they envisioned as a possibility in a land where freedom and liberty were the foundation upon which the new Nation would rest. The word equality, which was part of the

French revolution, was never part of our Constitution or Bill of Rights.

Ariana also realized that it was fortuitous that this subject had come up in her teaching schedule. She was very aware of the major game plan that Jonathan had shared with all of his students. The Ortho-Para Research Center was a huge complex that would eventually incorporate all of the acceptable modes of treating functional and structural illness due either to intrinsic or extrinsic causes. If successful, it would be the model for similar centers to crop up throughout the country and it would design the means of integrating the various modalities of therapy that had proved their worth in this first Research Center of its kind. The gathering of this proof was one of the goals of the Research Center and Ariana felt that to accomplish this end Jonathan and eventually Kenneth, who was obviously the one being groomed to inherit his position, would need some legal help in designing the Systemic Laws and Corporate Structure capable of controlling any egocentric factions designed to undermine the goals of the Center. Ariana was well aware of the factionalism that was ever existent in the medical profession, well before all of the Alternative Care healers had gained a degree of public acceptance. What would now happen, regarding intra-professional conflicts, if they are all put on the same campus and were asked to work in harmony when their very basal understanding of the human spirit and life was bounded by pure Reason on one side and pure Passion and egocentric intuition on the other?

Ariana smelled trouble in areas that Jonathan blithely ignored and Helen worried about, but chose not to bother Jonathan. She was working with Ariana to eventually create their own Constitutional Convention that included their class of 17, so that the fine, intricate points essential to preventing the undermining of the Goals of the Research Center could be written into a Declaration of Purpose that would be understood by every Healer who became part of the Research Center.

Ariana was hopeful that the factions entering the Research Center would not be as numerous as those that existed at the time of the Constitutional Convention in 1787. She also knew that Jonathan was less concerned and she feared that he would be made a dupe by one or more of the less acceptable Alternative Care healers who had little of worth upon which to stand but was compensated by the noise and vitriolic demands they designed to frighten. She knew that Jon was not easily frightened but if it could be prevented, it made sense.

These were the thoughts she carried with her to New York but made no mention of them to anyone. It was now time to purge herself of these concerns and Steve had proven himself a good listener when these needs arose at not infrequent intervals.

It was only the week before that Steve had done a soil study for Dan and had found a few artifacts, indicating the area of the Center was once inhabited by an Indian tribe. He was examining them when Ariana called him to the supper table.

"Steve, I know that I've harangued you about this before but the legal ramifications regarding Jonathan's project harass me and I need your solace. Listen to me and give me your opinion." She was not demanding and her sweet, endearing smile was all that was necessary to get a big hug of affection and a willingness, if not a sufferance, to once again be her audience.

"My ears are all yours darling and with the amount of law you've taught me over the years, I might have grown to the level of a knowledgeable critic."

Ariana laughed, "I agree with you darling. Knowledgeable, you've always been, whether it is the Laws of Nature, the Earth, or the Laws of Man. Here is my problem. How can we protect the Research facility, a system which will be global within 5-10 years, from becoming an Institution with men like Menzer who repeatedly tried to destroy Jonathan when he was at the Johnsonian Foundation? There must be a way of weeding out men and women whose greed and power needs are destructive to any system they enter." Her anger was dripping with frustration.

Steve responded "I don't think so babe. There are many men, bright and creative, who have lingering needs to assume roles of power but as you have told me many times, their dangerous ambitions often lurk behind a specious mask of zeal for some brilliant or righteous goals. Eventually, when the mask is lifted, the despotic intent that underlies their true ambitions are revealed. I don't think you can prevent them from becoming part of the Institute, but you should be able to get rid of them when their true despotic goals are revealed. See how easy it eventually was to get rid of Menzer. One man, the titular head of the Johnsonian Foundation, revealed Menzer for what he was. He did so in a public setting and Menzer was removed."

A smile came to Ariana's face. "That's right. I forgot about Menzer. Mr. Johnson made quick work of him, once he was convinced that his brilliance in Psychocybernetics had another heinous side that could destroy not only Jonathan, but undermine the entire Johnsonian Foundation. That's a good example that should capture Jonathan's attention and prevent him from poo-pooing any effort to awaken his need to be cautious and not too trusting."

Ariana was exuberant. "Thank you, Dr. Geologist, for solving my legal problem. Is there anything I can do to help you with some geological mystery?" She approached Steve and embraced his head and brought it to her lips. She whispered "I love you Dr. Steve, Earth Scientist magnificent, and my dear lover."

Steve was consumed and when he had recovered from her amorous attack he began laughing. "Yes my love. I do have a problem that you might be able to help me with. In the branch of Geology known as Petrology, there is a classification of rocks based on their origin. The first and most beautiful specimens

132

are the igneous. They own their origin to the liquid magma upon which the tegmental plates of the earth float. When this magma cools and dependent on its local mineral content, we have the various kinds of igneous rock.

The second type is the sedimentary rock which we could also call secondary rock. It is due to the breakdown of igneous rock into its constituent parts, and then in its muddy forms, due to whether it has broken down due to mechanical, chemical or organic means, and it eventually hardens into layers, they ultimately become what we call sedimentary rock.

The third type, which is the subject of my inquiry, is the metamorphic rocks. There are two types. The first are Ortho rocks, derived from igneous rocks, and the second are Para rocks, derived from sedimentary rocks. When either igneous or sedimentary rocks sink into the earth, in contact with the magmacious heat that created the rocks in the first place, they are rekilned into molecular aggregates and re-fired into a new and beautiful rebirth.

I don't know when this classification was created but I know it is before Jonathan's classification came into existence. If you look in the dictionary, Ortho is considered a primary prefix so that it fits well with igneous rocks which are also primary rocks. In contrast, Para is considered a secondary prefix, so that it also fits well with sedimentary which are made up of micro-granules of primary rock.

Ariana started laughing, "So what professor is the legal problem I'm supposed to solve?"

Steve remained straight faced with only a hint of a smile. "I was getting to that Ariana, but you have to be patient. Lawyers are not the only wordy professionals, a fact I am trying to teach you. You are aware that an Ortho, just as Yang, represents the qualities of masculinity, solidarity, firmness of action, primary substance, and at times a crystalline clarity in his views that seem to be an extension of Father Nature."

Ariana roared "Hey, wait there. The phrase is Mother Nature. Are you delegating Mother Nature to a Para role, a secondary role like Para rocks, which are the by-product of the wind and water breakdown of secondary deposits? What has that to do with Law and some significant problem you needed to solve? " Ariana knew now that she was being teased and was struggling to find a way to make it blow up in his face.

Steven was also struggling to find the reason why he started this subject in the first place. "Well Ariana, don't you find it strange that Ortho Hydrogen and Para Hydrogen also exist?"

Ariana burst out laughing again. "Now, smarty pants, I'm going to surprise you. Ortho Hydrogen is made up of two protons rotating in the same direction and, as is the problem with most Ortho dominant men, is somewhat unstable. On the other hand Para hydrogen has its protons rotating in opposite direction and is the only stable form of the molecule." Ariana was glowing with her performance and enjoyed the look on Steve's face as he was outmaneuvered in that encounter with his lovely wife.

"Well darling, I bow to my beautiful and most stable Para wife. Do you think the stability of Para hydrogen has anything to do with Alternative Care healing?" His face indicated he wasn't joking.

"In answer to your question Steve, it is 'Yes', and more than that I can't say."

Before they went to sleep Steve pulled out the paper that gave them their assignment for the month. They were given Humor therapy, Music therapy, Light therapy, Hypothermia, Hyperthermia, and Dance therapy.

Chapter 17

The class was not surprised when Jason, previously R.J. came to the meeting with Beth. She was not a member of the original class of 17, but she was first introduced to the group by Laura with whom she got her PhD in Traditional Psychotherapy and later by Helen who made her a part of the Ortho-Para team. It was through Laura that Beth met Jason, and the rest was history. Jason and Beth finally wed two years before the New York meeting.

Jason was the giant of his class and had now requested he no longer be called R.J. His new role as Professor of Physiology at Cornell University in New York and Director of Activities at the Ortho-Para Research Center, placed him in close contact with Beth and Laura who were the diagnostic team that referred patients for specific types of therapeutic activities based on a complete history and the Ortho-Para test (OPT).

To Helen, Jason, Laura and Beth were being groomed to be the clinical team that would ultimately bring Ortho-Para Medicine to the public, and Kenneth, who had his M.D., was now considered the final consultant for the more difficult clinical problems that came to the clinic.

It was for this reason that Jason's needs for therapeutic space would make him a primary adviser to Dan who was constantly re-creating structural designs in relation to the functional needs of each specific therapy.

Beth and Laura were involved with Touch diagnostics, but did not invade the territory of the M.D. Only patients who had been seen by their M.D. and who felt their problems appeared to have elements of autonomic extremes were sent to Helen who eventually turned them over to Laura and Beth for OPT evaluation. Eventually a Script conference was designed that was to create the protocols that had uniform, recurrent success and did not have to be watched as closely by Helen. The patient was then sent back to the family doctor with a detailed account of what diagnosis was made, the finalized Script that was advised, and the reasons why that particular Script protocol was most frequently successful.

When the New York meeting had ended, Beth and Jason drove to their Condo, which overlooked Central Park West, with the intent of finishing up their diagnostic reports and recommending the needed therapy.

They settled down in their parlor and quietly began attacking the charts that had to be completed. It was while Jason was ruminating over what he would write that Beth decided to pursue a problem that was bothering her.

"Jay, I want to talk. Can I interrupt your train of thought?" She was somewhat troubled and needed to ventilate.

Jason immediately put his pen down and turned to her, "Happily Beth. I was struggling with something that will suddenly come to me when I'm not pursuing it so intensely. What do you want to talk about?" He was all smiles and obviously not disturbed by the interruption.

Beth took a deep breath and began ventilating. "Jason, I watch you carefully design a day of movement and rest and most of the time you seem to face each problem with serenity, without too many moments of reactive tension. It is a place I'd like to be and despite all of my training, it eludes me. We are so different. I understand that gender designs those differences only to a degree, but if I remember the Ortho-Para theory correctly, as a woman I am supposed to be serene and Para and you are supposed to be a rampaging bull and Ortho." She revealed that strange mixture of smile and concern as she continued. "I've never shared these feelings with Laura, Helen or anyone else and since I'm so subjectively involved, I'm uncertain of the answers I give myself when I'm alone and am being invaded by this adrenergic charge."

Jason was not shocked by this revelation. One of the qualities that he fell in love with was the quickness of her mind, her energized movements, and the beauty and gracefulness of an ever moving body. He listened sympathetically to her concerns, but didn't answer immediately.

The quality of thoughtful and a momentary detachment irritated Beth. It was one of the very qualities she hated in herself and she was unable to accept it as a feeling reflex that would remain until she recognized its automaticity, and with this acceptance it would slowly disappear. Jason finally responded, "Darling, I don't know if I want you to change an iota from the position you're in and have been in since I met you. You'll note that I was slow in creating a permanent relationship with any woman before I met you and as you've seen, in the last five years that you've known me, I've been surrounded by beautiful girls who grew up to be magnificent women, yet none of them invaded my heart nor stimulated my perceptual appetites until you came along." He stopped to watch the smile of joy that radiated from Beth whose spirit was lifted by his expression of adoration.

He then continued, "Beth, as you know, our Ortho-Para polarity is designed by both our birth polarity and the polarity that is slowly designed by the reactive stresses that touch on us during the first 6-12 years of our lives. Although, as a newborn, you were like every baby, a pure Para flower, happy and vibrant when your basal needs were met, and furious when they were not. The balance between the Para tranquility and the Ortho frustrations slowly alter the autonomic mechanisms of the mid-brain nuclei that control our Ortho/Para ratio. Our very experiences in life are constantly tuning and retuning this ratio that ultimately controls our entire chemistry. Via this route it influences our passions, strengths, endurance and behavior. When I think of the differences of our childhood, I see Beth having had to weather a turbulent storm, while I, with my xy masculine chromosomes, floated in a gentle, warm, encompassing and loving sea throughout the unfolding process. It's not a gender aberrancy that makes you, you, and me, me. It's the pathos and bathos of life that re-templates the Ortho-Para-stat that influences our unfolding on a day to day basis.

There is absolutely nothing that I'm telling you that you don't know better than I, but you've tried to close your eyes to the past travails that influenced your unfolding and think that everything you've learned relates only to patients and not to yourself. Are you hearing me darling? Are you hearing me?"

Jason was talking to a beautiful woman in tears. It was a clear and understanding revelation that she had ruminated on in quiet for years but had refused to accept it. It was now time for Beth to accept what she was, understanding the process that got her where she was and in that way comprehending that any change will be related to the joys and pathos she and Jason design in the future.

Beth responded to this when the tears had stopped. "Jason, I hear you, understand you, and want you to have the remarkable woman you deserve. Thank you for being so understanding and I know that the energy I now feel rising is not my punishment but the wonderful generator that will be the fuel that energizes my life, marriage, career and the future that both of us will make happen, especially when we have our first baby. Jason I love you".

Beth went to him with tears of joy in her eyes and Jason held her, quietly stroking her hair. The silence and touch was the Para of love.

When they parted Beth began laughing, "You know Jay, I'm a strange one. I call you to supper to feed you and then monopolize your eating time with my tears and worries. Thank you for being you. You are my Para therapy and I'm beginning to understand how important you are to Laura, Helen and I. But before I elaborate on that, let's eat."

Beth got up to turn on some relaxing music and came back to the table happily relieved of a burden she had been carrying for a long time.

When they had finished supper, Jason invited Beth into the sitting room. He felt delighted for having successfully released Beth from the demons that had clung tenaciously to her for so long. He wanted to share with her the details of what he did with patients sent to him by the Ortho-Para group and many other specialty groups that which heard of the results of his therapy.

"Thanks for a delicious supper darling. We have both been so busy with our respective responsibilities, I never detailed for you exactly what I did after you, Laura and Helen gave me a diagnosis and asked me to implement a Script that you outlined. I only will outline my therapeutic responsibilities in relation to the Script. I'm sure that most of the non-mobile and non-mesophilic responsibilities encompassed in the Script, you are more aware of than I".

Beth smiled, "That may be true Jay but after seeing you handle your wife, I believe your physiologic expertise is much broader than you think and that you can handle areas of therapy never covered in your training in Sports Physiology."

Jason remained straight faced when he responded, "Thank God my ability to learn didn't stop when I got my PhD. Any contact with Jonathan and Helen added to my knowledge base and if it could be translated into a significant change in therapy, I use it. What I want to share with you can best be demonstrated with a case study.

We'll call her Jane Smith. She was seen by Helen ten years ago after a thorough physical exam by Kenneth revealed no easily diagnosed physical problem. Her complaint was a hyperkinesia, hyper-mobility and frequent bouts of projected fury. All of the hyperkinetic, organic, hormonal diseases, such as hyperthyroidism and aberrant chromaffin, metastatic disease had been ruled out. Everything was structurally normal and yet the ceaseless motion that didn't appear to be similar to Huntington's Chorea was ever present.

Helen took over and initially gave her the Ortho-Para test and her Ortho dominance was so extreme it had literally destroyed her relationship with all of the loving, protective and all encompassing systems of her life. She felt and was truly friendless.

An Adlerian study of her family constellation suggested that her mother and father hated each other and overtly and loudly spat out their derogatory language at their six children every waking hour when they were in contact with them. They belonged to no church, had no contact with a nurturing person and while in school, because of lack of training, she was so bad that even nurturing, loving teachers rejected her.

I was not surprised that she found her moments of balance by becoming a prostitute and saturating herself with love starved men who paid her well because she aggressively and happily satisfied their needs, at the same time fantasizing she was being loved and wanted to always sense her post-coital tranquility. It was only after getting a venereal disease that was resistant to therapy, that she sought Jon's and Helen's help.

Most of the Ortho-Para pathology does not fall into such extremes unless it's also associated with severe psychotic symptoms, with delusions, hallucinations and periods of detachment into another world. Jane may have been close to the breaking point but she hadn't crossed over the sanity line and Jon and Helen thought she was fixable.

They handed her over to me with simple instructions. 'Be her friend. Take nothing she says in anger personally. Be sure she swims one hour a day and if she doesn't know how to swim, teach her. Make a miler and then a marathon runner out of her and if necessary, run along side of her and watch for personality changes. When she appears exhausted with this routine she is to be water walking in the company of one of your very compassionate associates who can talk to her about different types of gentle touch therapies.'

At the end of two weeks of this therapy she was sleeping well, which was the first good sign that we were helping her find her Para side. Her therapy was exhausting for me but she found in me a man who only offered her kindness and company without any demands other than the exercises she could perform and without hearing the debasing criticisms which had always been part of her life.

Jane Smith is no longer a patient. She is one of Laura's friends now and I hope one day you have a chance to meet her. She no longer lives in this community and with her change in personality she soon found a male friend, married him and has become an ideal Mom.

For the sake of brevity I made this a short tale with a happy ending. It's not always that way because it takes an enormous amount of time and patience to care for someone who is exposed to environmental verbal poisons during most of their infancy and childhood."

Jason was through but ready to answer any of Beth's questions.

She, in turn, was tired and happy and was ready to detach from their work. "I can think of many questions darling but not this evening. Before we put our work aside, I think we should look at the list of Alternative Cares we have to inquire about."

Jason took out a small memo book and shared with Beth the healers that they were required to visit: Spiritual therapy, Sports therapy, Different forms of Massage.

Wallace Salzman

Chapter 18

From the moment John learned that the water molecule followed the same Laws of Systems and Particles as did Man, his love affair with water began. In Jonathan's first Treatise, he tossed accolades at David Eisenberg and Walter Kauzman who wrote 'The Structure and Properties of Water'. It became the molecular bible for John to study and now, at 35 years of age, he was considered the new authority on this complex water molecule and its multi-systemic functions, essential to every living cell- flora and fauna.

John's fascination with water and George's fascination with oceanography were the natural circumstances that brought them closer than any of the other male students in the class, and it was George's wife Cassy who introduced John to Judy. It proved to be an explosive romance. Judy was a sea photographer whose love affair with water was ultimately the reason that brought her in contact with Cassy who was working the emergency room at the University Hospital.

Judy had been commissioned by Aqua Magazine to take stills and mobile shots of a reef that was dying off of Key Largo. Her Scuba tank was new, much lighter than the one she initially used and she found she needed heavier weights around her waist to maintain the neutral buoyancy she needed to swim down to the reef that was growing 200 feet below sea level.

It was a beautiful sunny day with the sun high in the sky and the seas, like an aquamarine lake, allowed visibility rarely seen on a deep dive. She was already 60 feet down when she realized she forgot her watch and depth gauge and to go back would take up too much good visibility time.

She therefore felt depth and time estimations would have to take their place, and she dove to a depth that required she use a water tight lighting system designed specifically for depth photography.

She was shocked to see a reef of pillar coral that looked like a primitive forest of petrified trees. It allowed the small fish to swim between their limbs, but any large predatory fish would be restricted from entering this small fish haven.

Judy became enchanted by the dance of the small fish and the beauty of the flowering coral that were in full bloom in the few coral that were still alive. She was intent on getting as many shots as possible before she surfaced.

She became suddenly aware that she was getting too relaxed and was alerted to the realization she had been down too long and that she had the incipient signs of Nitrogen narcosis that had sneaked up, like an anesthesia, on her nervous system.

She began to ascend in hopes that a slow ascent, and the slow equalizing of water pressure that would occur, would allow her to exhale the Nitrogen fast enough to clear her brain and avoid the narcosis of the deep that had taken the life of one of her teacher friends. She was also concerned that if she came up too fast she'd be inviting bends or a pneumothorax due to too high a pulmonary gas pressure. Her logic was impeccable but her timing was wrong and when she had risen to 66 feet she began to feel pain in her chest. At the same time she could tell that she was running out of air and though she knew she would have to dive deeper to get rid of the pain, without air she could not.

Judy would never forget that day. First, she had gone diving without a buddy. Second, she did not bring her depth gauge or watch. Third, she told no-one that she was going diving and was at the mercy of the young man who had taken her out to the reef in a small boat and remained in the boat until she surfaced again.

When she was brought into the emergency room, writhing in pain, Cassy quickly surmised what had happened after talking to the native rower, and she was brought to the large pulmonary hyperbaric chamber and immediately compressed with Cassy next to her side. It was nearly 30 minutes before Judy opened her eyes and smiled at Cassy. "I do believe I owe my life to you. I'm Judy, what's your name?"

"I'm Cassy. You're lucky the young man who brought you to shore, and then to the hospital, was able to give me a brief but accurate history. What, in God's name were you doing? You didn't have a buddy, a depth gauge or a watch. Have you ever had a damn lesson on the dangers of diving?"

Judy grinned as she responded. "I've been a master diver and teacher of Hyperbaric Medicine at your University. I'm just a damned fool who forgot to bring all of my equipment and thought I could get away with it. Did the diver retrieve my camera?" Judy couldn't remember anything other than the moment she felt tranquilized. Fortunately it was only after she got into the boat with all of her cameras that she went into coma.

This was the introduction of Judy to Cassy and eventually, via Cassy and George, to John.

When John left the meeting at the Plaza hotel he called up to the room where he and Judy were staying. "Judy, George and Cassy have already left the hotel. We have to be at LaGuardia in five hours if we're to be sure of our flight. I must be in Majorca on Monday and if we miss this flight to Barcelona today, we'll never make it."

Judy was excited. She was going back to the island she had first heard of when she read the biography of Chopin. This was again reinforced by an essay written by a Ben Yosif when he was young and who was trying to escape a marriage that had gone sour. Knowing she was now going to Majorca, she brought the article with her to read on the plane. She had yet not shared this with John so that when they were on the plane, relaxed and at 30,000 feet, she took out the article. John immediately saw the name of Ben Yosif.

"Judy, where did you get that article by Ben?

"In the travel section, under Majorca. Do you know Dr. Ben Yosif?" She looked at him in utter surprise.

"Know him? I've been immersed in the history of Ben Yosif who was Jonathan Fentonowsky's teacher. Judy, did you forget the name of the Research Center? It will be 'The Ben Yosif Ortho-Para Research Institute'. The very money that's putting it up has initially come from Nathan Ben Yosif, his grandson."

John was on a high. It was unbelievable that they were going to the island haunt of Dr. Ben Yosif. "Judy, how long is the article? Could you let me have a look at it?" John was like a child who couldn't wait to open his Christmas present.

Judy laughed at his enthusiasm and pulled out the article that was to be her introduction, once again, to Majorca. "John, relax. Sit back and I'll read you the first page that was written by an author who was in, I believe the Para mode you and Cassy are always talking about. Listen to this verbal music.

"Man has craved for the poetry of the infinite and the awesome solemnity of silence. It exists everywhere where he has striven to reach as near as possible to God. He has sought him out on the heights of mountains; within the very caverns of basalt of which mountains are made; and on the edge of beetling cliffs overlooking precipitous drops into the sea. He has found him everywhere but nowhere more surely than on a huge rock, half European and half African that juts out from the Mediterranean sea, just south of Spain, known as Majorca.

In contrast, there are a few land masses that reach out to the turquoise bays and there we find mounts or mountains of sand. It is there that we see life, creation and growth. These soft, gentle areas of tranquility seem to grow right out of the sea. The shells and fish come closer to the shore. Vegetations are bathed by the changing tides. The roots of Mangrove trees grow out to the water, just as their trunks reach up to the sky to bring nourishment and life back into the soil. The roots, the grass and all other living plants bind the soil into a new whole that man can plant his feet on, so that he too can grow roots.

Where the sea attacks the land, we have death and destruction. Where the land invades the sea, we have creation, integration and life".

When Judy stopped reading, they sat there looking at each other in utter silence. She eventually touched John's warm hand and rubbed gently. "John, did you ever meet this man?"

John smiled and thought for a while. "Yes I did Judy. It was at Jonathan's wedding and then again at Jonathan's graduation from Medical School. He was the major speaker at my graduation and it was just before he retired. Once again, it was at his funeral that I was present to honor him, along with his students, but I never met his family.

Obviously he was a man with many likes and although the Ortho-Para theory is the brain child of Jonathan, you can see in Dr. Ben Yosif's writings the Ortho sea tearing apart volcanic mountains, and then the Para sea creating life. Judy, you should be able to take fantastic pictures showing the extremes of life and death by merely observing the sea's behavior in Majorca.

I've been asked to solve a serious water pressure problem at a hydroelectric plant situated in Palma, the capital. During one week's stay we will have a condo at the Cala d'Or, a beautiful bay area with a magnificent sandy beach. I'll have plenty of free time to enjoy sight seeing while we're there and we'll have plenty of time to perform our Alternative Care responsibilities when we get home. Judy, do you remember which ones were assigned to us?"

Judy frowned, "Dear, I left the list at home but I think I remember them because they're so strange: Aikido, Alexander technique, Aston patterning, Cell therapy, Chanting therapy, Color therapy.

Chapter 19

Sam, Jonathan, John and Justin were the only four men who were alone at the Saturday special meeting. They and their wives realized that this particular gathering was primarily for the class of 17 and their wives were happily free to shop while their husbands listened to Jonathan's presentation.

During the early grade-school years there was little confusion between young Jonathan and his teacher because at that time teacher Jonathan was known as John Fenton. When John, the teacher, became Jonathan Fentonowsky, his birth name, the confusion between the two Jonathan's became a minor problem that lasted only moments.

Although the men sat together at the morning meeting, they each left individually to meet their wives at designated locations. The day was beautiful with a delightful breeze coming out of the Gardens of Central Park. Sam left with a wave to his friends and caught a cab to the Metropolitan Museum of Art.

There is little evidence, during Sam's early years when he was in grade-school, that he was going into law. He was not overly verbal and aggressive, the type who you would think would go in that direction. In contrast, Ariana showed every quality of open-ness and argumentativeness that one would expect in any of the law professionals. Sam, however, was relatively quiet so that when he told Jonathan that he was going into law and then politics, he and the rest of the class were astonished.

Sam was now 35 and was considered a prophetic political philosopher, the guru who was frequently visited by the Representatives of both parties in Congress and the Senate because of his profound accuracy in predicting the economic wisdoms of a particular moment in time, a fact that was confirmed again and again by Nathan Ben Yosif who had been his closest friend for years after they both left the State University.

At first Nathan went on to Medical School, but his contact with Sam and the profound advice he was given on an investment strategy, made him entranced with the market and his consistent huge profits. Eventually it was Sam who enlightened Nathan regarding his grandfather and convinced him that if he was ready to get to understand his grandfather, whom he hadn't ever seen, it was most likely to happen if he got in touch with Jonathan and Ben Yosif's significant other, Helene.

Most of Sam's expertise was cloaked in a strange mystery that was never the subject of open conversation even amongst the many powerful political figures who sought his advice. Each individual considered Sam their private treasure and chose not to share his wisdom with anyone else.

This was the position that gave Sam the greatest comfort. To find himself behind a curtain of protection, unexposed to the liberal press and the paparazzi, so that he could comfortably have a private life without exposing his wife Amanda, a magnificent artist, and an up and coming power of the Board of the Metropolitan Museum of Art.

Both Sam and Amanda had gone through their professional crises. The resistances that they had weathered and to which they both responded, precipitated the essential growth appropriate to their respective fields. In terms of autonomic balance, Sam was a powerful Ortho-Para, with far less anxieties than when he was younger and single, and Amanda was a creative Para-Ortho, only after she overcame those moments of depression and hysteria that plagued her alone moments when she was in her early twenties. Their serendipitous meeting was at the office of Dr. Helen Fentonowsky whose advice they both sought at the time of their emotional crises. Amanda was sitting in the corner of the waiting room with eyes partially closed, detached from the agitated stranger who paced the room restlessly and relentlessly. A loose rug was located fairly close to Amanda's chair and as Sam was approaching Amanda, he caught his foot on the carpet edge and began falling. His struggle for balance had him tilting in the direction of Amanda who was startled and grabbed out to protect herself against his falling body. His struggle was in vain and he fell to his knees in front of her chair and when he looked up, utterly embarrassed, he found himself staring into the eyes of an angel who was now steadying his shoulders and laughing hysterically over this accident of fate. He too began laughing and just remained on his knees in front of her, feeling better moment by moment because of the beneficial effect of the laughter to both of their emotional extremes.

The noise was heard by Laura who rushed out of her office and was shocked to see her classmate Sam in a position of reverence, as if he were proposing to this beautiful stranger.

On that momentous day Amanda was seen by Laura and Sam by Dr. Helen. After a thorough OPT exam they were both autonomically classified and Sam was not surprised to hear that he was Ortho-Para with a 4/6 Para/Ortho ratio that explained his hyperactivity. Amanda, in turn, had a P/O ratio of 6/4, placing her in the Para-Ortho polarity that explained her bouts of depression.

Their happenstance meeting, which was the therapeutic miracle in the waiting room, was the beginning of their romance and quickly led to marriage. This was 5 years prior to the New York conference. They rapidly realized that their contrasting personalities were the therapeutic miracles that they both needed and with Laura's and Helen's guidance they learned to read the signals of extremes when they were building up in either one, and quickly learned what was necessary to counterbalance these symptoms.

When Sam entered the Museum, it was different from what he expected and it took him a brief moment to realize that he had imaginedthe Metropolitan was the same as the Guggenheim Museum. The large hall led in many directions and with the help of a guard he found his way to the Administrative offices where he was to meet Amanda.

As soon as he walked into the office he was met by a bubbly Amanda.

"Hi darling, I don't believe I ever introduced you to Julie Philips. She is the present head of the museum board. Did I ever tell you about Julie?

Sam smiled generously, "You have, many times darling, but I'm thrilled to meet you Mrs. Philips. Everyone has heard of your Nobel Laureate husband and whenever Jonathan begins ruminating on his past, Larry Philips is always one of the main characters. I've also heard a great deal about your art, Mrs. Philips, but I never had an opportunity to see any of it."

Julie was now in her mid fifty's. Her artistic career in the last ten years had catapulted her to fame, so that while her husband had climbed the mountain of Science and Reason, Julie had done the same on the neighboring mountain of Art and Passion, and in so doing the tumultuous marriage of nearly 25 years had become the epitome of harmony and mutual respect and in the privacy of intimacy, a continual rebirth of love and affection.

It was from this mature and gracious position that Julie held out her hand to greet Sam. "It is so wonderful to meet you Sam. Amanda has been part of my life for years now and has been talking about you with great pride. I believe I know you well, even though we've just met." She then went over to him and embraced him, while revealing a broad loving smile. She was still a magnificent beauty and its revelation to Sam was a moment he would not soon forget.

Sam had now been introduced to another Ortho-Para miracle marriage, not dissimilar to his own. Julie was a powerful Para-Ortho, just as Amanda, and Larry was a super-powerful Ortho-Para, possibly more so than himself but the test of time and the trials of early adjustment had lead to a full and very complete life of Living. These words came up again and again and represented the two properties essential to a happy and well balanced relationship.

Julie was now all smiles. "Sam, I already asked Amanda and she referred me to you. I'm having lunch with Larry in the Museum dining room in a few minutes. Would you and Amanda join us? I don't believe either of you have met my husband Larry and I know he would like to meet you." Her beautiful face emphasized the sincerity of her request.

Sam was honored, "Mrs. Philips, I'd love to meet the one man whom Jonathan claimed was always his biggest intellectual challenge and to whom he bowed in reverence."

Julie responded laughing, "Sam, Larry says the same about Jonathan. Their mutual love for each other and the many personal sacrifices they've made for each other are legendary, but I'll let Larry talk. Come with me, both of you and we'll go for lunch directly."

When they walked into the Museum dining area, Larry was sitting off in a corner that seemed very isolated from the rest of the restaurant. When he saw the two girls plus Sam approaching, he stood up to meet them with a gracious grin on his face. Larry was the same age as Jonathan but there was a distinct contrast in their appearance. Whereas Jonathan was losing his hair, Larry was nearly completely bald. Jonathan was not fat but carried about 180 lbs to his 5'10". Larry was starkly thin and although the same height as Jonathan, could have weighed no more than 145 lbs. He greeted them joyously. "Hi Julie and Amanda, I see you brought a new guest today and I'm going to guess it's the brilliant Sam whom I've met on several occasions of celebration." He held out his hand in greeting.

Sam happily responded, "I believe we met at your and Jonathan's graduation from Medical School, Helen and Jonathan's wedding and at Bertha's funeral, three blessed events that will stick in my memory until Alzheimer strikes. I feel privileged to meet you again Dr. Philips."

Larry responded, "On the contrary Sam, while I descend this mountain others must climb and you are on your way up. Whenever I'm involved with my colleagues they ask me about Jonathan and the magnificent seventeen. Jonathan called me last night and said he was revealing one of his new Eurekas at a conference at the Plaza. Did he shock you Sam?"

Sam laughed, "Dr. Philips, he shocked all seventeen of us and then sent us off to do some homework to help Dan out with his architectural burdens."

Philips looked surprised, "I know it's a huge challenge Dan has been given and I'm not surprised he's been sharing the burden with all of you. As far as Jonathan is concerned, he regularly comes up with these amazing Eurekas. He used to define the difference between the both of us. He said I was an accumulative Intellectual. I stored all of the knowledge that was of worth and threw out the garbage. In contrast he was a concerned creative Intellectual. He stored all of the questions that had yet to be solved. Between the two of us, he claimed during a moment of modest egocentricity, there were neither solutions nor problems that hadn't at one time surfaced in our minds and conversations.

Sam, did you read Jonathan's Trilogy?"

Sam responded quickly, "I certainly did Dr. Philips and I understand why it's not the best seller, but I wish it was. If it could affect the behavior of parents in relation to their parental responsibilities, I think society would make a quick movement away from the concept that life is supposed to be a great joy ride, always seeking pleasure and fun, even if you have to drink and drug your way into that position. The problem is created worse, for once you've become dependent on this route, your growth stops and the magnificent joys of a creative life disappear.

Let me change the subject for a moment. I would love to see Mrs. Philips and Amanda's art, if it's on display this month."

Larry quickly injected his enthusiasm, "Yes, I agree. You girls have to give us a personal showing."

Julie was delighted to see Larry relaxed and ready to reach for his Para side. Amanda was equally enthusiastic and during the remainder of that day they were a bubbly, youthful foursome spending the day and evening together and getting what would one day be a useful relationship for all four.

When they reached home and were slipping under the covers, Amanda took a piece of paper and put it on the end table. She turned to Sam, "In the morning I'll read you the list of six Alternative Cares that were given me. I'm unfamiliar with them all."

Chapter 20

The Biogenetic Corporation is located five miles from the site that the Ben Yosif's Research Center will be located. There must have been a powerful and passionate push-pull that had directed Justin toward the ministry initially and then dramatically, after two years of an initial experience, to switch to Bioengineering. It required that he go back to college for post-graduate training. Jonathan was initially leery of Justin's move toward the ministry but felt it was better that he experience the feedback of this decision personally rather than be talked out of it by any senior.

Justin was now working at the Biogenetic Corporation and chose as his project the development of a Carbon chip technology that might ultimately replace the Silicon chip and increase the potential of an organ control mechanism that was more compatible with the Carbon chemistry of organic life. He was aware of the ganglionic structures that participated in the Ortho and Para controls of all organs and how there was an increase in kinetic function with Ortho dominance and an increase in regenerative and reconstructive functions during Para dominance. His project related to a major question. What if he could attach a Carbon chip to one or more organs that could monitor and either enhance or depress the functions of any of the major organs of the body? It was this physiologic dream that drove him into the basic research laboratories of the new Biogenetic Corporation and it remained unknown to Jonathan, Helen and the rest of his class, because there was still too much work to be done.

There was however one person, the previous Ginger Angellina, who was now his wife and confident, whom he met four years previously and married only one year ago. It was she who shared his daily progress toward making the new Buckyball Carbon molecule, potentially the most remarkable addition to controlling the create/destroy ratio of the vital individual organs of the body. Justin was the only one in the class who focused on the Micro and Neuroanatomy of the Ortho-Para mechanisms of human balance, so that even as he worked within the bounds of bioengineering, he had to study the mechanisms of the biologic creation and destruction of the various organs in the body. Needless to say the field was complicated and he needed a Para mate who was knowledgeable in the gross and Microbiologic sciences to whom he might talk and share.

Ginger had majored in Bio-physiologic Cybernetics. She was at the University when Justin had returned for his Bioengineering Degree. They met in a Bio-cybernetic class and became study mates and eventually marital mates for life. Any Eureka that imploded into existence was due to the combined effect of both, although she remained in the academic environment and stayed at the University, while Justin broke away from teaching and remained focused on his anxiety producing, complex dreams and his creative reconstructive passions with Ginger, the love of his life. Whereas Justin had no previous romantic interludes, Ginger had, during her teens, been the love object of many young suitors, but the magic of trust that is one of the qualities essential to a permanent bond did not come into her life until Justin arrived.

When Justin left the Plaza conference center he went up to his room where Ginger was waiting. He entered his suite with a cheery greeting, "Ginger, are you ready to enjoy a weekend of Para pleasure?"

Ginger was delighted, "You know, my love that is a question that has only one answer. We have two days to detach from our work and I want to do everything that makes me love you just a teeny bit more so that you're ready to be molested when I attack you sometime later on today."

Justin laughed, "That sounds like a challenge I'm ready to meet. Let's change to less formal attire and go to a place I've heard so much about and have never seen, Coney Island."

Ginger clapped her hands in joy, "Wonderful. I even know how to get there by train. It'll be fun and we'll really get to know the notorious New Yorkers.

As they sat on the BMT elevated train, high above the street traffic of Brooklyn, they had an excellent view of the ghettos, the parks and the newly developed elite and elegant areas of one of the five Boroughs of New York.

It took one hour of traveling before they could see the wide expanse of the Atlantic in the distance. By the time they reached Coney Island the train was packed, with more people standing and packed tightly in nearly unmovable positions. When the door opened at the final stop the train emptied, nearly explosively, and Justin and Ginger found themselves in a world they had never seen before.

Racial and social uniformity no longer existed. The world of diversity, in all of its wonder, was the miracle of most of the cities that were bound on one side by the huge mass of land that was 3000 miles wide, and on the other, by a huge and occasionally tumultuous ocean that was also 3000+ miles wide. The life of those who lived on this edge was the result of population hybridization, a mixture of people from every country and culture in the world.

It was this cultural flow that Justin and Ginger became a part of, as they walked on the Board Walk that stood above a huge sandy beach and continued for miles until they reached a community of homes known as Sea Gate. This community bordered a small bay that eventually entered the huge New York Port Authority and the inlet and outlet for the commercial life support systems of not only New York, but a large part of the U.S.A.

Their awe, as they walked, was soon interrupted by hunger as the smell of Nathan's hot dogs wafted deliciously by in the cool sea breeze and they turned their attention toward pushing through the huge crowd of patrons who were milling outside of the stand, everyone either eating or seeking the assuagement of their sea stimulated appetite.

As they were eating, Ginger turned to Justin "What did Jonathan lecture on that was so important and different that it wasn't part of Friday night's meeting?"

Justin smiled and was slow in answering because of the huge bite he had just taken. He laughed when he was finally able to talk, "Jonathan had a Eureka that he wanted to tell the class, but no one else. The long meeting at the Johnsonian Foundation had many people who were scientific competitors in the field of research similar to Jonathan's, and he was not ready to share his thoughts with them. What he shared with us was very significant and I'll share it with you at home because it is the revealing of the human unfolding process, and Coney Island is the wrong ambience for scientific revelations. But his Eureka will help us in our evaluation of the Alternative Cares we've been given to evaluate.

I wrote them down and have them here in my pocket. They are: Detoxification, Drumming, Feldenkrais, Heller work, Herbal therapy, Hypnosis and imagery.

Wallace Salzman

Chapter 21

When Jonathan walked into the Johnsonian Center, he was greeted by his classmates, at first with shock and surprise and then with an enthusiasm given only when one discovers a long lost brother. For reasons unknown to his 16 classmates, Jonathan did not go to his teacher's wedding and communicated with no one during his college years. He had given no reason to anyone and remained lost to his classmates until he read of Bertha's death in the obituary column of a New York newspaper and came to her funeral.

He sat in the last row of the funeral home, still away and isolated from his classmates and left immediately when the services were over and was therefore seen by no-one, including Jonathan and Helen. As far as everyone was concerned, Jonathan, the student, was still a lost soul.

On seeing Jonathan and Helen, an awareness of the losses he had self imposed began to weaken his resolve to remain detached from his past life and one week after the funeral he unburdened his soul while sitting with his teachers, Jonathan and Helen, at their temporary offices in Long Island. At first he was very anxious in their presence but that didn't last long because of the joyous greetings he received from them both who were delighted for the reappearance of one of their stars.

Dr. Jonathan, the teacher, looked at Jonathan, the student and he quickly went over to him and embraced him, "Jonathan, I shan't ask you what may have detached you from the group. All that is relevant is that you've returned and I'm grateful that you have."

Jonathan was a bit more relaxed after this joyous greeting but he had come to his teacher as he would to a confessional and until he spoke his piece he would not be able to relax.

"Drs. Jonathan and Helen, thank you for fitting me into your schedule. I can't tell you how many times in the past 19 years I've wanted to make this visit and was fearful you wouldn't understand what forced me to detach myself from everyone and forge a new path, unattached to the past which had suddenly become the pain of my life. I'm not here today to ask you to understand my actions and my silence. I want to share the dynamics of these past 10 years and then decide whether I am welcome to come back to the group I so abruptly left without explanation."

The moment he hesitated Jon interjected a comment, "You don't have to explain anything Jonathan. We have all had life crises and I guess you're entitled to yours. The important thing is that you want to return and you're welcome, unconditionally."

Young Jonathan smiled at his teacher and continued, "Thank you Dr. Fentonowsky. I appreciate your uncomplicated acceptance but I'm here to wipe the past clean by the power of confession, so please play the role of the priest and priestess and then forgive me if you can.

During my senior year at high school my dad became enamored with his secretary who was 15 years younger than he and a beautiful seductress.

I don't know how long this sexual affair lasted but by the time my mom became aware of it she was showing symptoms of an Immune Deficiency Disease. My dad immediately became suspicious, because he was also developing similar symptoms that drove mom to the doctor in the first place. To make the story short they both had AIDS which he had acquired from his secretary and then gave to my Mom.

It was no different than being hit by a bomb. The emotional relationship between Mom and Dad deteriorated to Mom's hate and Dad's guilt and it was in this state that they remained for five years until their death. I'm sure that had they gotten some Psychologic help and had allowed forgiveness to enter their lives, their resistances to the fungal pneumonias that took their lives might have been greater. But Mom was so devastated and humiliated that she isolated herself from all of us, including her aged mother and died a recluse one year after Dad died.

I, in turn, withdrew from everyone of worth in my life. I quit UCLA in my third year and went to Australia where I drank and bummed around for two years. I then came back to California and during the completion of my degree I met Joanne who was the catalyst who changed my life.

With her help I got a post-graduate degree in Architecture and Land Development while she was getting a degree in Interior Design and via our educations we became the ideal you taught us about, so many years ago. We had become Bipolar and Unipolar mates.

You see Jon, I kept up with your Ortho-Para works and understand the differences between the two bonds that make up a marital relationship. It was Joanne who was sitting with me at Bertha's funeral and saw all of the people who were part of my early life before the tragedy of my parents. It was she who convinced me to come to both of you and share my family tragedy and hope that I'd be welcomed back into the fold."

Jonathan and Helen were amazed that he was able to share with anyone this inner emotional war that he had endured. Jon responded with loving concern. "Jonathan, your ability to climb above this tragedy in your life makes you unique amongst your classmates who have yet to learn how decimated life can become when illness strikes and tests one's capacity to weather tragedy and grow with it, or else detach from life and living and regress to a pitiful and never forgiving child. Believe me Jonathan, there are too many people who spend their end-stage of life blaming the world and God for the tragedies of life and forget that as long as there is life, they have the power of turning tragedy into a painful, growth precipitating Eureka."

Jon got up from his chair and went over to his student. "I believe your Eureka began when Joanne came into your life. Is she out in the waiting room?"

"No, I didn't bring her, but she'll be coming to the conference at the Johnsonian Center."

Helen entered the conversation, "Wonderful Jonathan. We'll be anxious to meet her. From the looks of you, you're as gorgeous as ever and it must mean you have a good woman taking care of you."

When Jonathan left the meeting at the conference center at the Plaza, Joanne was waiting for him. They quickly got into their car and decided to go to a small restaurant on Long Island before they investigated the site where Dan was to build the Research Center. Although Dan, Dr. Jonathan, and Helen were unaware of it, Joanne and her Jonathan intended to play a significant role in design and construction of the Ben Yosif Ortho Para Research Center.

Part 3
Alternative Care Therapies

Chapter 22

Dr. Jonathan Fentonowsky was on the verge of justifying the reasons for the existence of many of the Alternative therapies now offered by Healer Practitioners, and was thus placing himself in direct antipathy to the vast majority of Traditional Medical Practitioners. There were already many young physician renegades who had adopted many of the Herbal and Body-Mind techniques being offered by those healers, but their understanding of the basic reason why they worked was still a mystery. The mystery was hopefully soon to be solved by Jonathan's new theories and in the process, a revolution in therapeutic techniques that was already slowly being accepted by an observant public, was now to find their justification.

Battles were not new to Jonathan, but this was different. Throughout his years of battling for the acceptance of his Ortho-Para theory, his main focus was on the ways it would improve the understanding and treatment of all patients, and it did not allude to the benefits potentially afforded by any of the competitive Healer practitioners who were rapidly growing in number all over the country and were beginning to seriously encroach on the Traditional medical delivery system.

Helen was very sensitive to Jonathan's moods. Since the Johnsonian Conference, he was very aware of the dangerous path he was taking. He was sitting on a chaise lounge in deep thought when she approached and spoke.

"Darling, I've never seen you so concerned for such a prolonged period of time. How is your present challenge so different from that faced in the last ten years? I don't understand your inordinate worries."

He looked up at his adoring wife, "Darling, I remember reading how prior to

1945 the physician was keenly aware that there was a power in Healing Touch. Throughout the millennia it was the primary magic that empowered most charismatic physicians to successfully treat most illnesses, and this was despite the absence of antibiotics and the invasive techniques of present day medicine. What is amazing is that despite these shortcomings, the physician, so limited, was still worshipped as a servant of God and the life forces that He represented.

With the advent of antibiotics and the new technologies that came into use, the methods of teaching physicians changed. After World War I, a large population of physicians from Germany, France, and England came to this country. With the advent of persecution in Germany, during the 30's and 40's, some of the finest teachers in every specialty immigrated to our own Universities. The teaching of chemistry, physics, physiology, and anatomy became the dominant classroom method of teaching Medicine and it hastened the withdrawal of the time necessary for bedside teaching. Helen, it was bedside teaching that taught the touch aspects of medicine. It was the young inexperienced apprentice, who watched the emotional interplay between the physician and the patient, who was learning that the creation of an intimate relationship was the most powerful tool used for bringing the dying back to life."

Jonathan stopped for a moment and then continued, "Helen, I want you to call a mini-conference. I want you to invite the young doctors from our class of 17, plus Dr. Si Shu and his wife. I need to prepare myself for the backlash that I'm not secure enough to deal with at this moment, so that I can respond to the pressures intelligently."

Helen was elated over this decision. "Jon, I'm so happy you made this decision. I think it's essential that you begin to understand the conservative traditional practitioners who have demeaned and degraded every form of therapy they were not taught in Medical School. They represent the largest contingent of physicians and you must be prepared for an onslaught, more challenging than any you've experienced at the Johnsonian Foundation. This time you are not only protecting Ortho-Para Medicine. You'll be seen as an advocate of every variation on the theme of healing that filled the vacuum created by the teachers of doctors at most of the Medical Schools."

Jonathan was in complete agreement. "Helen, would you call Kenneth and Ally, Cassy and George, and Si and Nia Shu. Let them know that I want a full day of their time so that a Saturday or Sunday would be best as far as all schedules are concerned."

Helen was surprised, "Are you sure you want that much time, dear?"

Her look of surprise and consternation provoked his answer.

"Helen, I'm very concerned about the Traditional Medical response in this country. Already 40% of medical care is going to these various Alternative Healers, not because patients think they're better, but primarily because they feel they're being treated by professional healers who are more concerned about their welfare than are doctors, and therefore give them the time they feel they need from a caring physician.

Our present project will be an overt justification for some of their techniques and will add an increased level of concern by doctors because the 40% of the population now going to healers will undoubtedly increase to 50%. I'm also concerned that people who now go to physicians who are doing a great job will start going to Alternative care healers who are not capable of taking care of the serious and complicated pathologies that only a physician should treat. I feel I'm walking into a trap filled with malicious vipers and I want to avoid it, if possible. Helen, this is not the same as the introduction of the Ortho-Para Analysis, which only added an important new dimension to human evaluation. This is touching on the inadequacies and livelihoods of thousands of medical professionals and it must be done with the appropriate concern for everyone involved. I want to do it right the first time and not be backtracking and apologizing to the world for the rest of my practice life."

It was very apparent to Helen that this would be a much bigger potential disaster than she ever realized. "O.K. Jon, I think you're doing the right thing. I'll make the calls right away and try to make the date for soon, so that we've faced this problem before the next class meeting."

It was two weeks later that Kenneth and Ally arrived at 8 A.M., prepared for an arduous and searching day. Within a twenty minute period the other guests entered the home, all dressed in casual clothes and ready for a probing search. Prior to the meeting, the general topics that were to be discussed had already been outlined and they all knew that although Traditional Medicine was to be scrutinized, it's essential worth must never be degraded.

When they had all settled in their large family room, Jon rose to address his guests. "Thank you all for giving me this time so that we can probe each other's minds and in so doing help me find peace in this Ortho-Para project. You're here because I've developed a paranoia. It is regarding the potential negative responses of the entire medical profession when they become aware of the powers they've lost to the healers of today, and who only usurped what the doctors gave up 50-60 years ago.

I'd like Dr. Shu to begin today's sessions with a discussion of the medical practice prior to the development of antibiotics in the middle of the 20th century. Shu, I'd like you to approach this anyway you choose, but it should mainly relate to medicine in this country. We're all aware that the science of Medicine was practiced in Europe probably 100 years before it began to be taught in American Universities. But after 1910, following the Flexner report from Johns Hopkins, we quickly began to catch up. I'll shut up now and I'll leave the rest to you, Shu."

Everyone was smiling as Jonathan struggled with his own verbal hyperplasia. Shu quickly took the lead. "When Jonathan called me last week and presented me with this assignment, I had to review an area of medical history that I had done once before when I had decided that something was missing from American Medicine. This occurred when I had just graduated from Harvard Medical School, which was comparable to the two other prominent Medical Schools in this country, Johns Hopkins and the University of Chicago.

This feeling that something was missing was probably because of my exposure as a child to the practice of Chinese medicine. What it was, I didn't know. But it was such a strong feeling I decided to matriculate at the Beijing School of Medicine and find what was missing. When I got my second degree in Chinese Medicine, the vacuum was filled and at least, to my knowledge, I felt like a more complete doctor than those colleagues who I met during my daily hospital rounds."

Kenneth interrupted, "I had the same experience, only for a moment. It was immediately cured by my knowledge of Ortho-Para medicine. I bet it is for the same reason that Chinese Medicine solved your problem." He looked about him and realized he had inappropriately interrupted Shu. He quickly turned to Shu, embarrassed over what he had done. "I'm sorry I interrupted, Dr. Shu. Please excuse me."

"I understand Kenneth. You quickly picked up on the Yin-Yang component of my Chinese training and recognized its similarity to the Para-Ortho training that you had. You're absolutely right, but let me go on with my assignment. It is apparent that although London, Edinburgh, Berlin and Paris were teaching the sciences essential to the foundation of a medical school, until 1847 even the two most elite schools in this country, Yale and Harvard, did not teach this scientific base and primarily taught Theology and Philosophy.

Significantly, in 1793 a change took place in the Parisian medical schools in France. It was proposed, 'That which was up to now lacking in the School of Medicine, namely the art of observation at the patient's bedside, would now become one of the main parts of teaching'. These reforms were firmly implemented and replaced the older facilities of the then didactic training in the classroom. With the institution of this apprentice-like system, Paris became the world's leading medical center and students from Europe and the United States turned to her great teaching hospitals for training and clinical experience.

From this era on, a Parisian doctor by the name of P.C.A. Louis had the greatest influence on the development of Medicine in this country. He was the teacher of Osler, who once stated 'The brightest minds among the doctors fell under Louis' influence, and they, more than any other impetus, influenced the scientific study of Medicine in the United States'. Oliver Wendell Holmes was one of his students and his comments became a powerful voice in influencing the changes that rapidly took place in the U.S.

Unfortunately the number of Medical Schools that grew detached from this progressive University and Hospital philosophy were many and they became the base of a large number of diploma mills that admitted students, without requiring a pre-medical college education. By 1910 there were 160 medical schools turning out physicians with less than adequate training. The clarion call for change brought Abraham Flexner, a prominent teacher, into the picture.

The advances in American Medicine were dramatically accelerated with the Philanthropic gifts made by Andrew Carnegie and John D. Rockefeller. The men who guided the distribution of these funds were Frederick T. Gates and Abraham Flexner, a business man priest and a formidable teacher. It was under their guidance that the prototype of the Modern University-based Research Institute was created. The first was named the Rockefeller Institute for Medical Research. By 1953 it was the world's foremost center for Medical Research.

Flexner traveled to Europe and was primarily influenced by the German University system and their emphasis on the basic sciences. In 1910, after a thorough study of the medical mills in this country, he recommended changes in the medical school curriculum that were instrumental in closing 80 of the 160 schools by 1948. However there was a significant price that this German influenced policy was going to pay. The French emphasis on bedside clinical training was weakened and the beginning of the battle between a classroom and bed-side curriculum began.

The most powerful statement made against the classroom methodology came from the mouth of William Osler. He stated, 'Cabined, cribbed, confined within the four walls of a hospital, practicing the fugitive and cloistered virtues of a clinical monk, how shall he forsooth train men of a race, the dust or heat of which he knows nothing of and this is a possibility, cares less? I cannot imagine anything more subversive to the highest ideals of a clinical medical school than to have over young men, who are to be our best practitioners, a group of teachers who are ex officio out of touch with the conditions under which these young men will live.' "

Jonathan quickly interrupted Shu to give him a necessary rest. "Shu, I see you approaching our concerns that began so insidiously in the distant past. We were seeing one of the effects of how System and Particle Law # 5 was beginning to take place. With the Philanthropic subsidizations that Flexner and his brilliantly trained foreign physicians were implementing, science in the Universities blossomed explosively but the price that was being paid was still invisible to most. Osler, at that time, was a practicing, bed-side trained physician who participated in the dramas of his patient's lives and was able to reflect this very well in his opinion of those who would attempt to teach that which they were not living. This didactic, classroom philosophy, that guided the Medical schools at that time, despite the cries of Osler and Cushing, created a relentless evolving process in which this classroom technique, taught by the salaried, clinical physicians, took over the very essential clinical years of the medical student's education. Had the Parisian instead of the German model been used, the loss of bedside clinical contact with patients would not have precipitated the changes in Medicine as quickly. But change was inevitable, especially after World War II, when taxation made more and more federal moneys available for use by the federal government. With this new power, the feds took over the chore of further changing the Medical System. I believe Kenneth that you are prepared to share with us what happened to Medicine after World War II. I want you to speak both to the gains and the losses."

Kenneth had been sitting restlessly listening to a summary of material that had been part of his past studies many years before. He was quick to pick up on the story.

"There was one thought that Dr. Shu emphasized which I want to focus on because it will set the tone for what will continue to happen, with few people being aware of the significant cause". He looked about him and could easily read their knowledgeable faces, indicating that they all realized what was coming next. "What could possibly be the losses that were difficult to see because of the growth of the ground breaking science that was coming out of the laboratories and medical schools all over the world? For one, the amount of time the doctor spent with the patient was less and the doctor-patient bonding time was replaced by history questioneers. We'll be talking about the tragedy of the bonding time loss and the trust that eventually went along with it. With the passage of time and the setting up of new methods of history taking, the young physicians, who were still struggling with the huge quantity of information that was daily being discovered, were unaware of the bonding loss and that which went along with it. Information was expanding, and new techniques of book-keeping for keeping this information were developed. Computer technologies suddenly invaded the doctor's offices. New personnel, with new talents, had to be hired as paramedicals so that within the office setting new stars were being created by the expanding environment.

It was in the middle of the 20th century that the Hill Burton Bill was passed by the Congress. It immediately subsidized a hospital building frenzy throughout the country. The Federal justification for this expansion was the need for increased availability for medical services. It naturally followed that Medical School subsidization would be necessary to increase the number of physicians graduated in order to service those hospitals and the growing populations about them. The expansion of the population was associated with an increase in the number of cities requiring all the services that population concentration required. Law # 6 was demonstrating its power of creating new stars within any expanding system.

Long before 1960 and primarily within the large cities, physician specialization began and the General Practitioner was witnessing the first danger to the concept of the wholistic medicine that he practiced. By 1960 residencies in specific specialties, instead of rotating internships, became the post-graduate route that the doctors took when graduating from Medical School. It was apparently the goal of the educators and the teachers to create Partistic physicians and limit the number of Wholists, the General Practitioners that would be created. As rapidly as the tree of Medicine sprouted branches and they in turn sprouted twigs, a new specialty came into existence. One surmises that the teaching profession felt that a man's brain, with finite association pathways, could no longer cope with the enormous growth of medical theory and medical technology.

Soon, with the continued growth of specialization and the concomitant growth of expensive instrumentation requiring special and tireless attention for skilled perfection, specialization became the growing force in Medicine.

With the concept of specialization, developing out of apparent need, there could not be a valid criticism from any medical mind who was first thinking of the health of Man, which Medicine was meant to serve.

Sadly when you think technically, a rational process, you forget about some other profound affects that were soon to be found by the critics of Medicine. As soon as you limit the responsibilities of man and narrow his horizons and his necessary scope of accomplishments, he has time for more moments of self evaluation, self justification and unfortunately, self deification. No longer needing to know all that there was to know about the body, the specialist elevated his particular role in life as a more informed man about a part, to a level of virtue higher than the man who was still waddling in the miasma of the whole.

The concept of the Partist specialist had begun as a product of need and became the cause of the schism in Medicine that would change the timeless image of respectability that had grown since the days of Aristotle, to one of suspicion and to some of scorn. Contrary to Reason's role in life, instead of harmonizing the pre-rational passions that the Partists and Wholists quietly felt about each other, passion was used solely as a weapon of destruction. The intra-professional strife was not camouflaged. The image of the doctor was being tarnished and the power of the specialty organizations rapidly changed the number of physicians going into Family Practice. This was 1960. How did Ben Yosif handle that period of intra-professional stress, Jonathan?"

"I remember him talking about it on rare occasion. I think it was before the General Practitioners decided to initiate the establishment of a Board of Family Practitioners that Ben maintained that a complete doctor required a dialogue between doctor and patient and required more than just a brief contact and interchange of information. He stated that a physician must stand in relation to his patients and must eventually establish an "I and Thou" relationship before the fullness of the intellectual and spiritual encounter can be realized. He felt you cannot succeed in treating the many aspects of a man's illness until you see him as something other than many defective body parts with need for biochemical or surgical readjustments. Reason, computers, or questioneer evaluators, could not take the place of the powerful spirit of a man who plays the role of friend and confessor and at the same time represents the magical unknowns of knowledge which give the physician the wisdom to reopen the dramas of life or close the curtain forever more. He does not represent a God, as many have accused the egocentric M.D., but certainly when involved in a patient's last illness, he is a close servant of God. When Ben Yosif was practicing, it was not the specialty oriented approach to patients as it is today. When someone was ill, he placed his total self, not a part of himself, under the care of a physician."

Jonathan now gave the floor to Allison. She began, "This story now approaches the Medicine of today. It has become a professional orthodoxy. We select our high priests, listen to their laws, and judge ourselves on the basis of their standards of excellence. The very relativity of Medicine and its short lived absolutes, should have taught us all a long time ago that caution is essential if we are to flower uniquely as creative, contributing individuals, and not become just another hot house flower fed the same knowledge and implementing the same advice as every other similarly shaped and highly predetermined physician.

Most of the medical changes and advances today have been paramedical and technical. They have been perceptual and not conceptual. Sophisticated instruments capable of revealing non-invasively and invasively the internal structures of our body are being used by physicians who are not responsible for their creation. Drugs discovered and synthesized by Research Pharmacologists, and Radiation devices discovered by Biophysical Engineers, have all been developed to assist the physician in improving his diagnostic and therapeutic skills, but the creativity comes not from the minds of practicing M.D's, but from the servant professionals who change laboratory theory into the practical tools essential to the practice of good medicine today.

While this radical revolution, in uncovering the details of human structure have taken place, human function, both individual, social and spiritual, remain attached to the same old conceptual scaffolds. The physician's understanding of the human condition has become as perverse as the consensus Endophilic society that is insidiously taking over our nation's value systems. Doctors are so busy rehashing the old concepts of Man and Behavior, there has been little time, nor encouragement, to discover a new base from which to build. Thus human functions, both internal processes and external behavior, have taken a back seat to human structure and because of this we discover more and more of the structural deficits associated with poor health and remain on a less stable base in our comprehension of the functional organ interrelationships that indicate good health within the human system or for that matter any system within which there is behavioral interaction."

Allison had finished what Jonathan had asked her to review and he then turned to Cassy. "Cassy we've had a rather brief critical review of Medicine up to 1960. What effect did this expansion have on Medicine from that point on?

Cassy approached the history of Medical change no differently than Ally. "Thank you Jonathan. The technology that followed on the footsteps of the 'knowledge revolution' developed so fast that the average hospital found it difficult to keep up with the latest and most effective CAT, MRI, and later PET scans, each costing millions of dollars and requiring frequent use by the hospital technicians in order to make it financially feasible.

If the rapid growth appears to be exaggerated, you must realize that there were outside forces making this technologic expansion an absolute necessity. The rise of the Intensive Care and Coronary Care facilities and the instrumentation developed to serve the needs of the deathly ill was the catalyst that catapulted the rise of the social protectors of our Society, the Lawyers.

Whereas the physician was the doctor who serviced the physiologic, Natural laws of the individual, the lawyer was the doctor of Social law who claimed to protect the patient from any violation of their rights that were protected by the Bill of Rights. Strangely enough when the physicians discovered ways of turning around what were once illnesses that invariably ended in death, they became more vulnerable not only to the lawyers but also to the patients themselves. The expectations of physicians grew out of proportion to their new skills, and nearly every physician discovered he was not invulnerable. With an increase in malpractice suits there was an increase in lawyers. Insurance companies and premiums eventually led to an overwhelming increase in physician overhead and the costs that every patient had to pay. I know that Jonathan is now paying over 50,000 dollars a year for coverage and he does no surgery and is not involved in invasive Medicine. Fifty years ago when Ben Yosif began his practice, his premiums per year were only 24 dollars and he did everything.

In summary, the problems of Medicine became the problems of the patients and with 40% of the population unable to get insurance, the doctors had not only made themselves too expensive, they had lost most of the magic that gave them the power to do what appeared impossible to the laity.

The magic I'm talking about was creative Touch, Love, Compassion and Trust. They had become technologists. The non-invasive Healers of Alternative Care Medicine were inexpensive; gave of their time freely; were willing to give the amount of time necessary to create an intimate, bonded relationship so necessary to build trust; offered simple services that were Touch rather than Tech oriented; and were clearly not dangerous or frightening, because they did not violate the integrity of the body. With growing evidence of their success, the Alternative cares have continued to increase and now represent a loss to Medicine that never had to happen.

What can we professionals teach our patients as to what criteria they must now use to avoid making a lethal error in their choice of professionals when they are not well? That was Jonathan's assignment to me, and it took me some time to come up with the essential criteria for not making a fatal decision.

I referred to the literature being published by the young doctors who are now being trained by physicians who represent the way Integrative Medicine is being practiced today. They've incorporated the practices of many non-physician Healers, who practice the Alternative techniques that are being offered as substitutes for some of the more invasive techniques now being used by Traditional practitioners. In 'Spontaneous Healing' written by Andrew Weil, M.D., he gives the following two lists (page 225). It is a summary of what Traditional Allopathic Medicine can and cannot do for you.

Can:
1) Manage trauma better than any other system of medicine.
2) Diagnose and treat many medical and surgical emergencies.
3) Treat acute bacterial infections with antibiotics.
4) Treat some parasitic and fungal infections.
5) Prevent many infectious diseases by immunization.
6) Diagnose complex medical problems.
7) Replace damaged hips and knees.
8) Get good results with cosmetic and reconstructive surgery.
9) Diagnose and correct hormonal deficiencies.

Cannot:
1) Treat viral infections.
2) Cure most chronic degenerative diseases.
3) Effectively manage most kinds of mental illnesses.
4) Cure most forms of allergy or autoimmune disease.
5) Effectively manage psychosomatic illnesses.
6) Cure most forms of Cancer.

Seek help from an unconventional doctor for a condition that conventional medicine cannot treat. Do not rely on an Alternative provider for a condition that conventional medicine can manage well.

Jon, I hope that this is what you were looking for."

Jonathan was pleased with the entire exposition by his students, in particular with the division of responsibilities that Cassy added to the presentations. He smilingly turned to Dr. Shu and challenged him.

"Shu, does anything strike you when you look at the 'Cans and Cannots' that Cassy presented to us?"

Dr. Shu smiled as he began answering a question that fit right into his concepts of medicine as taught in Beijing. "Yes dear Jonathan. I am immediately aware of one thing. The list of 'Cans' are therapies that would most likely be treated by Ortho techniques. The list of 'Cannots' would most successfully be treated by Para techniques. The question then arises, why is this true? Ortho illnesses, or I might add Yang illnesses, usually arise because of the patient's contact with the outside world. It is because of the contact with outside invaders, whether they be human, animal, bacterial, parasitic or the technologies of man.

Para illnesses, or Yin illnesses, come from within and just as is the case with the unknown that exist in every contracted system, they radiate little information regarding cause and lend themselves more frequently to techniques that encourage the body to heal itself and not to the therapies that come from without. In summary, Ortho techniques produce healing from without, and Para techniques produce healing from within. Jon, is that what you were looking for?" With a big smile Shu turned to Jon who was thrilled with Shu's insight.

"Yes my dear Shu. That is exactly what I was looking for and I'm beginning to see the light. I am however, a little uncomfortable with any absolutes and the recommendations listed by Cassy are a little too black and white. I am not ready to agree with Dr. Weil and his listed Cans and Cannots. In reading Weil, I see him floating in areas of uncertainty and it may be the only place we can be at this time because uncertainty still remains a very big part of Medicine and may always be. I hope that in our meeting next month, one or two of the presentations focus on this uncertainty or explains why it may be irrelevant to the functional illnesses that have their causes due to internally induced, rather than externally induced, pathology. Until our next meeting let's do a lot of deep thinking.

Chapter 23

Dr. Jonathan Fentonowsky stood smiling before the privileged 17 and their wives. One month had passed since their last meeting and a dissertation on a goodly number of the Alternative healing techniques now offered to patients, was to be presented by each couple with the goal of understanding the validity of the techniques and the amount and type of space that would have to be made available in the new Ortho-Para Center.

When mutual greetings had been exchanged, Jonathan spoke. "It's time to begin this meeting and I believe Kenneth and Allison were scheduled as the first presenters. How do you intend to make the presentation Ken, together or separately?

Kenneth quickly rose "Jon, we thought we'd divide the presentation between the two of us and make separate statements and then finish by summarizing our findings together and then responding to any questions before we're done. Does that meet with your approval?

Jon smiled, "Of course Ken. I think that format is a good way of squeezing as much out of you as possible because I want this to be the final meeting before we start on the architectural drawings of the Alternative section of the Institute."

Kenneth smiled and turned to his classmates. "Ally and I were given 6 topics to discuss and I thought I would begin with Reflexology because it is a take off on ancient Acupuncture and Shiatsu and maintains that in applying pressure and manipulating key points throughout the body, it would cause progressive relaxation. It was initially called Zone Therapy and came from Europe in the early 1960's.

When reading the literature on Reflexology, I faced the same difficulties I had when studying Acupuncture and Acupressure and even when I finally resolved my problems with both, I still had trouble with Reflexology. I decided to see a Reflexologist recommended to me by a patient who adored her.

Sharon was a beautiful woman in her mid-thirty's who greeted me as a long lost friend. I didn't play games with her and let her know that I was a physician who wanted to first experience her techniques and to then understand how she thought the therapy worked.

The first thirty minutes were spent with getting to know each other, not related to our professional interests but more to personal things such as our marriages, hobbies and worldly interests. As she talked she asked me to sit on a comfortable chair. She kneeled on a small rug before me and brought over a basin filled with a sweet smelling liquid. She then removed my shoes and socks and began to massage my feet and cleanse them at the same time.

As she continued, we chatted and I began to feel very relaxed and very spoiled. She then gave me a sheet and asked me to remove my shirt and pants, and pointed to a table for me to lie down. The foot massage then became a hand massage and eventually, with firm hands, she massaged my head, neck, upper and lower back and it appeared that my entire body would eventually become part of her domain. Her touch was gentle and rhythmic, but at times with powerful hands she pressed on some of the hand nodules that were developing under my palmar fascia.

At times she appeared to press on different parts of my palms and soles, telling me they were points of connection between the somatic outside and the visceral organs inside my body and then laid out, with diagrams, the relation of specific areas of the palms and soles to the organs involved.

I must admit that this lovely lady, who smelled of lavender, working in an environment with her soft musical voice was slowly putting me into a trance that I'm sure was somewhat hypnotic and during the 1 hour session I was experiencing not only Reflexology, but Massage therapy, Aroma therapy, a mild Hypnotic trance and the wonderful Para feelings that I received only from one other person in my life, Allison, who I gave greater privileges to on the day we were married."

The class and Ally laughed at the implications and Kenneth then continued. "When I left Sharon's office she had achieved a magic change in my autonomic dominance. I had become a Para child, happy, relaxed, flushed and the low back pain, that I had from lifting a heavy can the day before, was gone. I met Sharon only once and she had become a close friend, very easy to adore. It didn't take long before the outside world and the responsibilities I faced, soon jolted me back to the problem solving mode, but I still felt wonderful.

My exposure involved more than Reflexology. It encompassed Aromatherapy, Massage and by her explanations as to how they all worked, a form of acupressure that concentrated on the hands and feet but in her hands it involved the entire body, except for the pelvis. That I still reserve for Ally."

Once again the class roared and Ally feigned disapproval which made everyone laugh even more.

"I want to show you a diagram of the hands and feet and show you what Reflexologists claim are the key points that relate to the different organ systems.

REFLEXOLOGY AREAS

Manually working specific areas on the feet is thought by proponents of reflexology to improve the function of corresponding organs and body parts. For more information, see page 24.

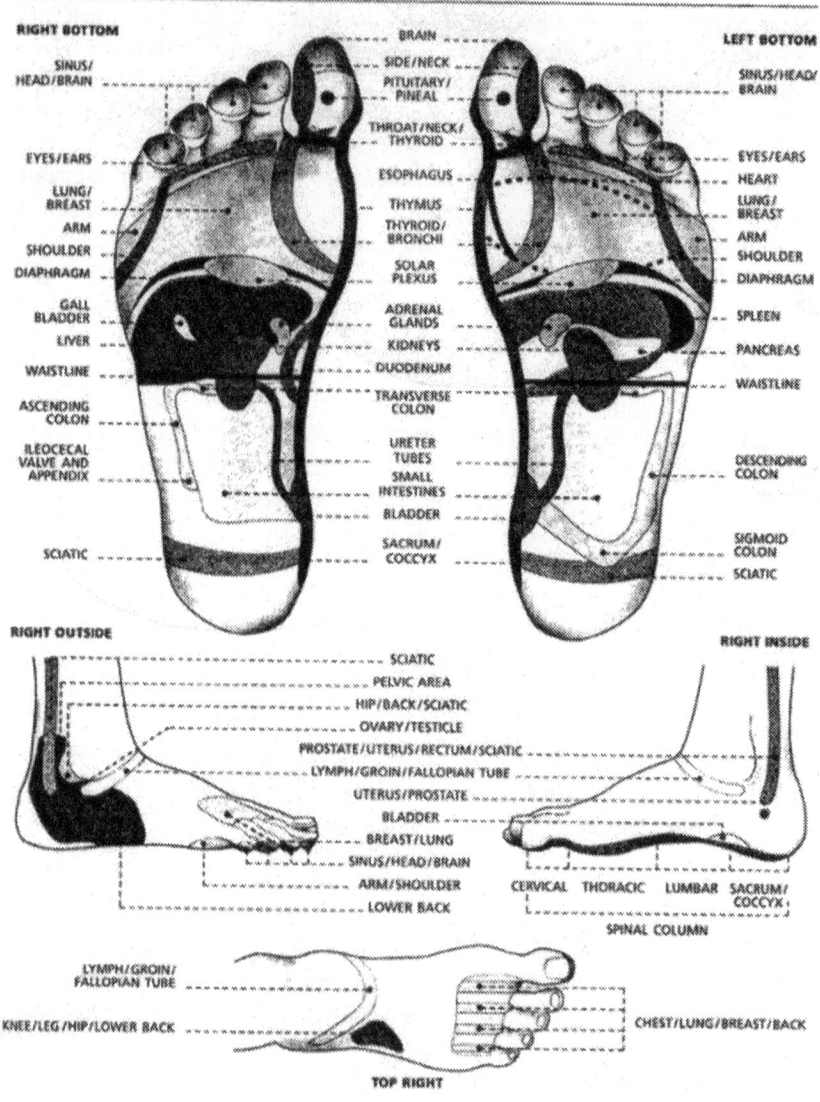

I want Ally to continue this conversation with a focus on Acupressure and Acupuncture and when she is finished we'll put it all together and try to make some sense out of it all."

Allison was laughing when she rose and took the floor. "I'd like to thank my illustrious husband for being aware of the total affect that intimate contact with his very private woman, has on him. I tell all you girls in the room that love and intimacy is not an Alternative therapy, but is a primary and essential therapy, especially for the Ortho individuals, their marriages and their families. From what you heard from Jonathan and Kenneth and I'm sure have experienced personally, the so called Alternative cares add to your lives a dimension of touch that was an available and essential part of every day of the first two years of your life. Living is a part of today and tomorrow. It is Ortho. Loving is a part of today and yesterday. It is Para. I can remember Helen telling me a story about her courtship with Jonathan. Early in their relationship he was primarily mentally focused on his lectures to students and ignored her to the point that she felt she was being misused. She withdrew and before long he came back to her, never to withdraw again. She was Para dominant and was 'Now' oriented. He was Ortho and was always solving the problems of the future. It was all summarized in his loving plea when they were making up after their arguments. He pleaded "If You'll Be My Today, I'll Be Your Tomorrow".

Everyone in the room sat quietly and listened to the story that happened 20 years before and remained the mortice that bound Jon and Helen forever. Jon and Helen just sat and smiled and reveled at the distant memory that was the first moment they knew their relationship was not casual, but permanent. Allison watched her teachers and continued.

"Today I am going to quickly merge the medical Alternatives that Kenneth has given me. Probably, the most accepted of the Alternative procedures, that a goodly number of traditional physicians have adopted, is Acupuncture. Historically it was preceded by Acupressure and has been practiced in the orient as part of Traditional Chinese Medicine for more than 3000 years. No different than Reflexology, which is relatively new, there are points on the exterior surface of the body that are considered to be attached by meridians and via them to every one of the visceral organs, so that when there is a failure in any organ Chi (energy flow), there is access to that organ via the external key points that can reestablish the energy flow (Chi) in the organ involved and in this way relieve the symptoms of the illness.

Anatomists have attempted to either corroborate a neural confirmation of this or debunk it as impossible, but the fact that it effectively has relieved the pain and symptoms of many conditions, that have been unsuccessfully treated by Allopathic physicians, stands as powerful subjective evidence that whether or not the theory as described holds true, the therapy frequently works.

There is evidence that Endorphins are secreted in large amounts with Acupuncture, but it then creates the question as to why the procedure at times gives sustained relief of pain and symptoms. There are those who speak of the Placebo effect, especially when treating pain that has its locus of origin in the brain, but whatever the dynamics of its actions there is one commonality with all the Acu-therapies, and that is a trust in the doctor and a belief in the therapy associated with a powerful desire to get well. For this reason Reflexology, Acupressure and Moxibustion are Para therapies which work on endogenously precipitated pathologies that may be Ortho or Para functional illnesses requiring Para and not Ortho therapies.

There are no absolute rules. Dr. Bernard Lowen, who designed and presented to the medical profession the concept of the Intensive Care Unit, described in his book "The Lost Art of Healing", a moment in his life when he injured his back and had all of the symptoms of low back nerve compression. He was in China evaluating Chinese Medicine with a group of multi-specialty physicians from Harvard and he immediately put himself in the care of American Orthopedic men who hospitalized him and put him in traction. It produced no relief, and after a few days Lowen asked to see a Chinese Acupuncturist. The hosting physicians called on an old man from a neighboring village to treat Dr. Lowen. The first attempt with the needle only increased the pain. The second attempt relieved him completely and he was able to walk again.

This is not an isolated story and it was done on a prominent physician who was as skeptical of the procedure as the average Allopathic physician is in this country."

Jonathan was concerned with the lengths of the presentations and interrupted Ally. "Ally, I think it's time for our illustrious audience to ask questions and I'll usurp the floor with the first question. How much space do these therapies require?"

Kenneth was quick to respond. "I think if Daniel has a joint waiting area with toilet facilities that are shared by all the offices, two rooms, one for history taking and personal hands on therapy and establishing a relationship with the patient, and the other, a large room, maybe 20 by 20 feet, that has available space for furniture, instrumentation, shared music, proper soft lighting, the number of such facilities would be only dependent on how many Alternatives will become part of the Alternative addition.

The private offices should be acoustically isolated, as should all private offices so that the privacy of the consultation area is not invaded by unwanted noise.

Chapter 24

Jonathan then turned his attention to Daniel and Becky. "Dan, did Kenny and Ally enlighten you with a possible solution to your design problem?"

Dan was all smiles. "If no other person adds functional information to the data I collected from you, Helen, Kenneth and Ally, I think I can work out the functional space when you decide on the number of people you want involved in the center. Understand that it isn't necessary to put all of the Alternatives that ultimately become part of the center in one building. We can design sites where architectural bonding can be practical, beautiful and functional at the beginning, and build additions later.

With the advances taking place in all branches of Medicine it is essential that even as we design the present, we must successfully anticipate the future.

Becky and I had some deep discussions regarding the Alternatives tossed our way and we went through a crisis moment that related to Becky's questioning of her own personal significance in a world that may become Godless."

Jonathan immediately stood up and stopped Dan from continuing.

"Dan, what do you mean a Godless world? Who took God out of the world?

I certainly didn't." Jonathan appeared furious because he felt his theory of the origin of Man's concept of God had been grossly misinterpreted. Dan quickly responded.

"Jonathan, please don't take this personally, but it's essential that we understand that there are many people who will declaim that your theory was a clever means of saying 'God never was. He was a creation of a two year old infant.' Understand Jon, Becky is the one who will be contending with the backlash tripped by your theory and we recognized the need for a response to the loud accusations that will come her way before they attack you personally."

Jonathan simmered down and finally relaxed. His initial reaction was a responsive reflex. He recognized the fears of gentle Becky who was not the warrior type capable of weathering an attack by her own congregation, and possibly all of the local clergy, whatever their religion or faith. He quickly apologized. "Danny and Becky, I should never have responded to your potential dilemma as if it were unlikely to happen and that is why I was insistent that the lecture be given in the seclusion of an inaccessible conference room, and that until you were given the OK by me, you would not reveal it to a soul. I still haven't decided how the statement is to be presented publicly but I want you to know that Becky will not only be protected as a servant of God, but the concept of God will be strengthened in any statement we make, so that it will be more difficult for the forces of power to use Him for their own personal power and to negate the power of Religion in this world.

Becky, you must remember that I said "In order to template an awareness of a Higher Power, God knew it had to be placed in the substructure of the brain. It had to be in a non verbal place and therefore not eradicable via reason's machinations. In truth, the ratiocinations of man could not be used to prove or disprove God's existence. All thinking men would have been imprinted at an early age with a passionate proof of His existence because it was imbedded in the brain as deeply and as inaccessible as breathing and all the other vital, essential reflexes that support life.

A child is created because of the power of love that God imbedded in man's reflexes and which would not flower until the functions of Mom were complete and the child matured into the position of awareness that included the need for a mate, a child, a saviour and finally God, the overseer of all creation.

Daniel and Rebecca, I want you to think back to 20 years ago and what you both, along with the class, decided how it all began with Mattergy, that undifferentiated energy created by God. This Mattergy had the inner potential for ultimately creating all that eventually came into existence and which would ultimately become part of our Cosmic future. You understood and believed that Science and Reason did not and could not answer all the questions that Man could conjure up and would undoubtedly ask throughout the history of mankind. From everything I've said, lectured on and wrote, was there any indication that God, Passion and Para were to be written off as a fantasy or fairy tale?"

Becky quickly responded "No Jon, but somehow I felt that my very spiritual foundation was being attacked." Becky hesitated a moment and then began to falter, so that Dan had to jump in and rescue her.

"Jon, please be patient with her. I feel it essential that you recognize that this is a very real problem that you will have to face and in the way you handle it, your very own Ortho-Para theory can be elevated significantly in the Medical, Theological and Spiritual spheres or because of a perceived attack on the very base of religion, be set back for eons of years."

By this time Helen was struggling to contain her pent up anxiety.

"Stop, all of you. I want you all to regain your calm and composure.

Dan was protecting Rebecca, who certainly could not be considered the representative of all humanity, but she represented a large percentage of them. Jonathan understood that this was an example of what we feared might happen in the real world and that it was best that it happened in our small world, so that we have a greater awareness of what we're up against. The first step is to discover exactly what tripped this reactive fright and depression."

By this time Becky was crying. The embarrassment before all of her classmates was too much and Dan was unable to calm her. Helen approached Becky and put her arms about her and talked very quietly, while the rest of her classmates sat still and watched. As Helen observed Becky's reactions to the obvious concerns of the class, she felt it best to remove Rebecca from center stage and led her out of the room to a nearby office.

"Becky, calm down. We must talk and we can't do it when you've lost control. Tell me darling are all of your feelings related solely due to Jonathan's theory? I truly believe that there is something else bothering you, something that you're afraid to share. Becky, am I right?"

Becky immediately lost the semblance of control she had slowly gained and she was sobbing uncontrollably once again. Helen realized she would have to jolt Becky into control and withdrew and sat on the other side of the room, leaving Rebecca alone and to herself.

The room remained quiet with only the sobs of Becky disturbing the silence that pervaded the room. Helen just sat and watched Becky from a distance as she slowly lifted her head and found that her support system had slipped away and she was now truly alone. The Minister, who was trained to be a support system for others who were grieving, suddenly came to her senses and stopped crying. She wiped her eyes and noseand walked over to Helen who showed no evidence of moving toward her and she suddenly awakened to the fact that she had been caught in a lie by the Clinical Professor of Psychology, Helen, and it was time to come clean.

Becky stood before Helen and whispered "You know what is happening?"

"I think I do Becky".

"How could you tell?"

"Because your emotional turmoil could not be a reaction to Jon's theory and every time Dan approached you, your reaction of grief became more profound. You're pregnant, aren't you?"

"Yes I am. I realized that when my period did not come on time and a pregnancy test was positive. Suddenly I realized that my whole life, my ministry, my marriage which is only two weeks away and my self esteem had been polluted by lust. I was just a slut and deserved not to be a mother or minister of

God."

Helen grabbed her and insisted Becky look at her without tears. "Becky, you're 35 years old. You are about to be married to a wonderful man who sired a child who will become your most precious accomplishment, when you've mothered him properly. One of the most important qualities for a wife and mother is compassion and sensitivity and you're overdosed with both. I'll bet you haven't told Dan because you think he'll think less of you and will want to terminate the relationship. Your hormones, as Jonathan would say, have screwed up your brain. I want you to remain here alone for a minute and I'll be right back."

Helen left, and Rebecca, just by confessing her guilt, had gotten control of herself and sat quietly waiting for Helen's return. It was only minutes before she wandered in with Dan.

Rebecca intuited what Helen was doing and stood up, completely unaware of what Danny would say but finally blurted out "Danny, I'm pregnant. I told you it would happen and it did."

At first Dan was startled and then with a smile he ran over to her, grabbed her, lifting her off the floor, "Becky, how wonderful. God that egg must have been waiting for that moment of love making and what a wonderful gift we now have in the making. Becky, I think it's wonderful. What do you think Helen?"

Helen had a broad smile on her face. "I think it's wonderful and the only problem I see, is going ahead with your marriage plans quickly so that your wonderful baby won't have to be too early. I knew that Becky's emotional breakdown was due to more than Jonathan's statements."

Helen approached the two and embraced them, pulling them close to her. "I want you to know that you're not the first couple in your class who got hitched just a short time after culminating a love affair. Remember the baby is a product of Natural law at work. It is God's law. The wedding is merely Social law at work. It is the State's way of taxing a perfectly beautiful manifestation of system contraction. Rebecca, Jonathan will be delighted and so will every one of your classmates who have all experienced the passion of conjugal love and the creative miracle that follows in 9 months. With both of your permission I'd like to announce this to the class and let everyone participate in the making of your wedding."

Becky turned to Dan who seemed elated and she burst into a joyous smile and said, "Come on Dan, let's do it."

They returned to the conference room startling all present when they saw the smile on Becky's, Dan's and Helen's faces. Dan waved to Helen and Becky to sit down and he went to the stage, eager to talk to a surprised class.

"Becky and I want to thank you all for your patience during this emotional moment in our lives. We both now understand why she so emotionally alerted

Jon to the possible problems he might encounter when he makes public his theory of how God performs the miracle of imprinting his presence within the minds of all mankind. It is no less a miracle than the flowering of a child, born with an empty but magnificent brain which ceaselessly imprints knowledge throughout life, under the heavenly guidance of a loving Mom. Rebecca is pregnant and will soon deliver another of God's children."

The class responded with screams of joy as they all rose to salute Becky who would soon take on the chores of responding to every wish their new baby would have. The love that the baby has for Becky will one day be transferred to that highest of powers we all know of as our God.

Chapter 25

Jonathan was all smiles following the revelation of Becky's pregnancy and accepted her criticisms regarding the misinterpretation of the meaning of his thesis, as a very valid one. He made serious note of the intensity of her response and once again rose to talk to the class.

"Cassy and George, I believe it's your turn to teach us the relevancy of Ayurvedic Medicine and in what way it relates to Traditional Chinese Medicine, Curanderismo and all the major religions in general. I realize that you probably have had the most difficult concepts to research, but with Cassy as part of the team and with her deep awareness of the complexities of Ortho-Para Medicine, I hope you were both able to struggle through the cultural base of these concepts so that you can see their relevance to the treatment of illness, especially illness of the mind."

Cassy turned to George "Do you mind if I usurp the podium and attempt to give depth and breadth to these wonderful faiths?"

George quickly replied "Cassy, until you get lost in the ocean depths, as an Oceanographer, I'm irrelevant. However I would like to make some relevant comments when you're through. You go ahead with our rather difficult assignment."

Cassy placed her reference notes on the table in front of her and began what she knew would be a long dissertation. "I realize our assignment was related to Alternative Medicine in this country, but it's difficult for me to call Traditional Chinese Medicine, Ayurvedic and Curanderismo Medicine forms of Alternative Medicine. They are medical theories which are thoroughly integrated into their religions and cultures and have many concepts that parallel those of Ortho-Para Medicine.

Ayurvedic concepts were popularized in the United States by the many books written by Dr. Deepak Chopra. At times, religion and culture attempt to encompass not only the human, but also the Natural laws that are the foundation of our structural physiology and the reflexes automatically designed to guarantee life and living.

In reading the text of the Bhagavita Gita, the Hindu bible with the commentaries of Sri Aurobindo injected into the interpretation, I am probably as close to an understanding of its message as anyone can be because of Ortho-Para theory. Aurobindo carefully pointed out that there are two approaches to the interpretation of the messages defined in the text. The older Orthodox school laid stress upon the spiritual, religious and other-worldly aspects of the goals of spiritual man. Although not the same, it seeks the same goals as those recommended by Chopra. The newer interpretation, as represented by Aurobindo, dynamizes the old, other-worldly, quietistic concepts and injects spirit into a work-a-day life, so that an immanent spirit participates in the immersed and transcendental ends that man ultimately reaches out for. In doing this Aurobindo has raised action completely out of the mental and moral plane and has given it an absolute spiritual life. Several modern philosophers have focused on this ultimate goal as evidence of a major evolutionary leap and in his latest book 'Sex, Ecology, Spirituality', a huge book on the Spirit of Evolution by Ken Wilber, we see a brilliant dissertation that reasserts the teachings of the Gita. It states that by immersing our self into the nothingness that the process of meditation seeks, we ultimately create the possibility of ascending into what he calls a transpersonal position in which one can ultimately become One with the ultimate Spirit of God." George quickly rose to add his comments to Cassy's last statement.

"It is this that I have difficulty with. The inner, selfish, survival pacemakers that are essential to the awareness of one's hungers are looked at as an undesirable part of Man's nature and elevates the soul, the transverbal and transpersonal spirit, to an evolved position that one should be seeking as one lives one's life. They call the natural, biologic maturation process of the infant and child and that part of man that should never be completely obliterated, a weakness that must be demeaned and replaced by this after-life dedication. One can tell that this document was written before man realized that he was free to control his own destiny and could reach for a position of self sufficiency, a position that didn't become possible until education became Universal in the fourteenth or fifteenth century A.D. The Gita sought the total ablation of the vital powers that were required, if we were to mature, and in a sense elevated the process of life's renunciation to spiritual and Godly levels rather than elevating the vital life of challenges and goals that ultimately led to fulfillment and self sufficiency. I see nothing but mental illnesses in a life that is designed, not for the fulfilling of one's life's needs, but in its place, the fulfilling of the spiritual demands of a God that says 'All that you do and think must be in dedication to me. You must be free of desire, a false force imposed on you by Nature which you must eventually recognize as merely a process of growth which binds you to the Earth and to earthly things, and therefore not permitting you the freedom to dedicate yourself to the pervading spirit, who is the ultimate perfection.' Everything that these fundamentalistic religions say creates the illnesses that Medicine ultimately has to treat. I may sound terrible, but I don't believe that a God, who can create the Cosmic miracles that we've discovered with the God given powers that lie within us, would require more than our thanks each time we discover another creative miracle, as we live a fruitful life here on this Earth. Jonathan, I'm not a non-believer, but I believe we were designed to live and fulfill our complete destiny on this wonderful earth and sea, and not spend our life seeking an after-life as the ultimate goal that so many Fundamentalistic Religions and Philosophies profess as the highest goal that man can attain.

I am not a physician, but from what I've seen of Man, Family and Nation, there are many functional illnesses related to human and system interrelation-ships that Ortho-Para Medicine must study with the cooperation of the other branches of the life sciences, in order to understand the countless functional diseases caused by the fixations of the mind related to the different phases of life and the possible and impossible goals that man reaches out for, at the wrong or right moments of maturation."

Cassy was once again approaching the podium and George spit out the last sentence very quickly because he knew his speaking time was over.

"George, I must comment a bit on the concerns that you registered. I don't think Aurobindo personally believed in the ancient Orthodoxy so rigidly imposed by the ascetics. It was the Vedists who represented the position in life where action and works could not lead one to a higher evolvement because the liberation of the soul (the goal of the ascetic), required the giving up of desire, life and work.

The Vedantics, that Aurobindo represented, stressed that evolvement to the ultimate Sattwic (ectoendophilic) position was possible, but to do so required that the individual pass through the Tamasic (Endophilic) phase, the period of egocentric, selfish indulgences and then on to the second phase, the Rajasic (Mesoendophilic) phase, emphasizing the growth of power and leadership and then ultimately to the phase of brilliance, the Sattwic, when one more con-sciously focused on guiding the soul toward detachment from one's worldly works and prepares oneself for the ultimate detachment of the soul, when death closes down the energy pacemakers that made the action and works no longer a necessary part of life. As in all religions, there appears to be a built in con-flict, an inner dialogue related to the needs of this life and the work needed for self sufficiency, and the meditatory requirements for achieving the gift of an after-life. This duality of life's purpose can be the seed of inner dissension and the Psychologic illnesses that make functioning in one's systems of responsi-bility very difficult. Needless to say the conflict between spending our life in prayer to the service of God, or to our multi-systemic responsibilities involved with serving our marriage, family, village, church, state, and our nation, creates an inordinate tension. The focused stresses can be patently constructive or destructive, and it is only the latter that can lead to Para and Ortho dominant illnesses.

For this reason, it was important to include Religion in our discussion of Alternative therapies. Without an understanding of the causes of destructive stresses of any kind, a full awareness of which activities require elimination and which are constructive and good for you, would never be understood."

Jonathan thought he would intercede at this time. "Cassy and George, that was an excellent introduction to an area of Wellness and Illness that is frequently not pursued by physicians because it has been allocated to the religious authorities and in the past was divorced from Medicine.

I admit that in my practice and with my focus on the autonomic imbalance between Ortho and Para, I spent very little time evaluating the stresses you mentioned, although it has, in the past, been the swamp of ignorance that has caused more wars, more deaths and more eternal hate and prejudice than any other cause of world chaos I've studied.

This is the case in Traditional Chinese Medicine, Curanderismo and in truth every other faith practiced in these United States and the rest of the world. What is essential to understand is that we are not born with a genetic propensity to harbor religious antipathies. They are taught and although I truly believe that a good physician is a teacher, the eradicating of the world prejudices is a job so enormous it can only be attacked by the clergy of all religious denominations when they themselves rise above the biases that create the vertical hierarchies that place everyone on a ladder of different significance, so that we can look up and see those who are better than us, and down, so we can see those who are to be distained.

Interestingly, the Vedic and Vedantic faiths became economically and religiously the base of the caste system in India. But it is not India alone. The economic power hierarchy exempts no country from this laddered caste system. Remember, the United States was one of the last countries in the world to give up slavery and to this date there is greater and greater evidence of the segregation of the poor from the rich; the weak from the powerful and a political hierarchy that is designed to maintain this power differential of the few over the many.

We may also have to remember that although the Hindu religions of India examine life and the hereafter in a slightly different way than the Judaic-Christian, the Islamic and Chinese theologies, the ultimate consequences of them all is the creation of a clergy that is subsidized by their parishioners to perform the priestly rituals that make believers more secure in their stressful journey through this life. In truth they are deific doctors, teachers of a faith that helps parishioners cope with the struggles of this life, and also crystallizes the concept of an after-life which finds its greatest significance in the aged who have had to come to terms with their mortality. They can then place their souls in the hands of the clergy and their bodies in the hands of the physician. They are both essential to pacify the ill and make the healing of the wounds of life and living more comfortable. Cassy and George, I wondered whether you felt there was a place for a Global Chapel in the Research Center?"

Cassy quickly responded to the question, "Absolutely Jonathan and I believe its Universal design should be the modern creative responsibility of Dan."

Dan quickly injected his feelings, "Cassy, I agree with you completely. After listening to both of your presentations and the conflict between a "this world" and "after-world" existence, I'm sure I can get some creative artists to assist me in the inner design, but I think I know what I'm doing with the outer structure."

Helen and Jon began laughing. Helen asked "Dan, you're not thinking of a crystalline roof over a pulpit, are you?"

Dan and everyone in the room began laughing as he answered, "I sure am and it's going to be magnificent."

Chapter 26

Jonathan approached Cassy and George. "I want to thank you both for undertaking the study of the Vedic and Vedantic concepts of one's life and their great relevance to Wellness and Illness. I think it's time to get off the transpersonal goals of the soul, since we are the keepers of the body-mind, and now hear from Laura and David who will focus on several Alternative therapies that are fascinating in their claims. Laura, why don't you start the ball rolling."

Laura, with her beautiful smile and startling red hair, faced her classmates and began. "It is fascinating that when one looks at the electronegative elemental series, the first product of Creation, Hydrogen, was right in the middle of the series and is as much electropositive as it is electronegative. In previous discussions, we defined electronegativity as the Para state of the Periodic table and electropositivity as the Ortho position. It is also fascinating that the basal element of all inorganic and organic matter, Hydrogen, represents the perfect internal balance and harmony between the Universe's expression of Ortho and Para, of male and female, of intrusion and incorporation and of positive and negative. Considering that both the electric and magnetic fields also have positive and negative poles, the concept of gender differential and social differential should be considered the natural consequences of Cosmic creation and gender neutrality, the unique position of Hydrogen, the mother element of all matter.

The intrusive phase of life is being actualized when we're awake and taking on the challenges and responsibilities of sustenance and survival in a not too giving outside world. To do it well, we attack. It is this phase of the day when we change ourselves into a kinetic energy creating machine and change our body structure (potential energy) into movement and via this route precipitate the magnetic, electrical and chemical changes that manufacture the ATP fuel used by all living cells.

The positive magnetic field of the sun enhances these dynamic processes, which are normally catalyzed by the serotonin producing Pineal gland. The incorporative and regenerative phases of man's daily cycle, the circadian rhythms, take place during the night. It is at this time that the production of serotonin availability comes to an end and melatonin, a by-product of serotonin, takes over the stimulation of the nocturnal reconstructive processes. There is then a reversal of the potential energy > kinetic energy processes when the kinetic energy and the catabolic by-products of the day before, act as the template for the reconstructing of potential energy in the form of structure. This process is in relationship to the resistances (physical, mental, emotional and spiritual) encountered during the preceding twelve hours of wakefulness.

The earth's negative magnetic field dominates during the night and is in contrast to the sun's positive field. The south's negative magnetic polarity dominates during the night and the north's positive magnetic field dominates during the day. They are the magnetic counterparts of melatonin and serotonin, of electronegativity and electropositivity, of incorporation and intrusion, of Para and Ortho, of Yin and Yang, and of feminine and masculine. They represent, in all of their actions, the collective attempt of every particle and system to seek out balance, either within themselves or along with another particle or system that contains the opposite polarity and in attaining this, an opportunity for a moment of balance.

The Ortho-Para ratio is a destroy/create ratio and exists in one way or another in not only organic life, but also in the processes that eventually alter all of matter. Having presented Ortho as representing the destruction and break down of structure, one could easily interpret this as being the ultimate and only cause of human and cellular death. This is no more true than representing Para as the creative polarity, that in dominance might be the source of eternal life. It is in understanding the ultimate need for both, of Ortho and Para balance, that we will understand the pathogenesis and patho-physiology of both Wellness and Illness and by so doing elevate the complex relationship between the physician and patient to one of teacher and student and in this way eliminating, as best one can, the rescuer-victim concept that came into existence when antibiotics and technologic medicine were born in the late 1940s and early 1950s.

In examining the above antitheses and referring to the first treatise, we can draw on other analogies. It is during contraction that the bonds of Para and intimacy are formed. During expansion and Ortho, these same bonds are broken. Since Ortho and Para are biologic extensions of the laws of Systems and Particles and therefore the corollary to the laws of contraction and expansion, we'll be able to use these laws as guides when we microscopically focus on the physical, chemical and physiologic processes that control, integrate and harmonize the temporal elements of Ortho and Para oscillation and in this way allow us to both sustain and recreate ourselves in relationship to the resistances encountered during the day. In this sense cellular living and dying are in constant battle so that we, the multi-cellular creatures called human, can have a sustained life for much longer than our cells.

To extend our analysis even farther, we must touch on the conflict between Science and Faith, Physics and Meta-physics, Reason and Passion, logic and intuition, conflict and peace, relational closeness and separation, Natural and Social laws and Yin and Yang, which is the Chinese counterpart of Para and Ortho. Note that there are threads that significantly fuse these particle and system oscillations to each other so that despite their meaningful differences they are inseparable."

David was fascinated with the in depth presentation of Laura. As a genetic scientist he was very familiar with genetic contraction and expansion since it was what he did in the genomic research labs every day. "I think Laura, it's time for me to reveal some of the clinical studies that you and I read about and then experience as we pursue the subtle powers of solid magnets on some simple structural abnormalities. But whatever we report should not surprise you. At a meeting we had some time ago, we spoke of layered man and demonstrated that as life got more complex it was with the adding of a new layer that had the power to alter, in some magical way, the ancient layers that lay beneath it and in this way complexities grew upon complexities. If we look at man closely, the Archeologic remnants of the primitive past still exist as part of his meta-cellular community.

Although my specialty has been in human biogenetics, the history of solid and electro-magnetics has always fascinated me. I'm surprised that Medicine has not been able to finally clarify the bioscience of magnetism so that its therapeutic benefits or dangers were no longer the seed of conflict between the medical charlatans and an ignorant and confused medical profession. There is certainly a reason for a thorough study to be done since magnetite was used by the Chinese and Greeks as long as 2500 years ago.

With this in mind, I decided to review the history of biologic research on the effect of magnetism on the human body. What amazes me as I reviewed the history of the biologic effect of magnetism on the human is its absolute absence after its first discovery by the Chinese. For that matter it wasn't until 1771 that Galvani wrote a paper on animal electricity and in 1883, Helmholtz wrote the first articles on the speed of nerve impulses and opened a new era of research on the relationship between electricity, magnetism and ultimately between electromagnetism and life.

Physicists in the late 19th century discovered that every electrical current or force has a comparable magnetic field, induced at right angles to the flow of the current. The very process of generating electricity was by the movement of a dynamo, a powerful magnet at right angles to an electrical coil.

It therefore amazes me that it wasn't until 1959 that Kolin was the first to magnetically stimulate a frogs sciatic nerve and in 1965, Bickford and Fremming stimulated human nerves with magnetic fields. Slowly, but surely, the relationship between magnetism, electricity and biologic life began to integrate itself so that that which was used by the ancients 2500 years ago was being reawakened and successfully creating a new concept that is still struggling to find its use today in Medicine.

By 1989 magneto-encephalograms became a way of examining brain function and at the same time the miraculous discovery of Magnetic Resonant Imaging demonstrated that we could manipulate the charged proton particles of the body, the hydrogen molecules of water, and via magnetic fields and electromagnetic wave induction currents, create an image of all the tissues of the body. Yet, despite all of this evidence of bio-magnetism and the history of its successful use by cultures so long ago, Medicine and its research leaders chose to ignore the past and in this way by-passed a possibly non invasive way of treating some of the ills of the body. As these facts unfolded, I realized that for the twenty years I've been a bio-scientist, not once, other than in the history of a physician by the name of Mesmer, who was declared a charlatan by his colleagues, did I hear of magnetism being used as a therapeutic modality.

Consider some simple facts. We are born into a world that has a magnetic field that varies between .4 and .6 gauss, dependent on how close we are to the earth's poles. These magnetic fields polarize the tissues of the body and are obviously physiologically essential. Man has been shown to be biomagnetic on a basis of animal research and the manipulatory powers of the MRI. This is not hocus pocus stuff.

It is here that we find that the science of electromagnetism, as described by physicists, far exceeds the science of bio-magnetism, so that the difference between the two sciences opens the door for unverifiable claims, anecdotal miracles and a level of skepticism that casts shadows on this new and growing field.

Let's first examine the measurable science that is unattached to marketing.

Perpendicular to any flow of charged particles is a magnetic field whose intensity is proportional to the intensity of the electrical flow. If a magnetic field is placed perpendicular to the flow of charged particles, it enhances the flow in relation to how it enhances the preexistent magnetic field. Electrical and magnetic fields are married to each other and can both enhance and depress each other accordingly. This data, that has been demonstrated in the field of physics, applies equally well to the flow of electrically charged particles anywhere in the body.

We are therefore talking about the flow of ions, the charged elemental particles in every channel where there is fluid flow. This includes the blood vessels, the lymph channels, the peri-cellular and intracellular fluids, the intestinal and bile fluids, the cerebral spinal fluid and that within every nerve and muscle.

There is a variability of flow, as we pass from the large arteries, arterioles and capillaries and then on via the venules to the large veins and vena cava. Each vessel has a magnetic field that lies at right angles to the flow of its blood, serum or lymph and any manipulation of this magnetic field leads to alterations of the fluid flow and the responses of the charged particles to both the north and south fields of the magnets. The north is electronegative and tends to attract positively charged ions such as sodium, potassium, calcium and magnesium. In turn, it repels chloride, bicarbonate and phosphate ions. The south is electropositive and has an action that is the exact opposite of the north.

There is a relationship between the effect of the magnetic field and the pH of the blood. The higher the concentration of the positive Hydrogen ions, and therefore the lower the pH, one can expect more intense reactions because of the greater number of ionically active particles. In this sense, during the day when we are creating higher amounts of kinetic energy, so that we can move and perform our responsible roles at work, the process of oxidation is breaking down body structure and turning it into energy. In this process, hydrogen is released from the contents of body structure and creates the acidosis and all of the toxic acidic by-products of structural breakdown, and therefore the low pH.

During sleep, a reverse process takes place. Via the processes of reduction, the Hydrogen once again becomes part of structure. The organic chemistry is not complicated. The process is catalyzed by tissue enzymes, and regeneration and healing takes place. With the loss of acid and hydrogen ions from the blood, serum and other body fluids, the blood becomes more alkaline and we are repaired and prepared for the next day's work.

Let me change our focus for a moment and return to my own expertise as a biophysicist and put aside electromagnetic physics. There is a create/destroy process going on in the body at all times. Body structure is either being created or destroyed. During the day, structure is being changed to kinetic energy and manifests itself in the way of movement. During the night, the movements and the toxic waste products, that are the by-product of the catabolic processes that create the energy, become the templates that control the nocturnal recreation of structure. The chemical template is the pattern that designs the restructuring of those tissues that have been stressed, hopefully a little more efficiently than the structure which had been functioning the day before. It is an endless dynamic in which the rest period, under the dominance of the Parasympathetic Nervous System, recreates, regenerates and reconstructs the cellular damage associated with the processes of living. Living our daily lives therefore destroys and kills the underlying cellular structures of the body. When we detach ourselves from the world in sleep, a form of functional dying stops and our cells begin to live again so that cellular living is a Parasympathetic function, and human living is a form of Orthosympathetic function and is the chemistry that dominates our bodies during the day. We call Parasympathetic, Para, and Orthosympathetic, Ortho.

It is apparent that a good sleep is a busy time. It is the time of unconscious RNA and DNA reshuffling. Just as the goings on in the brain that is hidden from our vision, the nocturnal recreation is only revealed when the new controls being integrated are suddenly implemented during the day.

The nocturnal reorganization of the human is always in relationship to the unresolved stresses of the day before. Yesterday has signaled the changes necessary for tomorrow and in our cerebral inter-phase, during sleep, we are being recreated in relationship to these failure of yesterday. Our competence is dependent on which Embryonal maturation level that is being stressed. It is in our persistence that we reveal to everyone those inner values of significance for which effort is worthwhile. Any intense effort produces slowly evolved improvements, along with a necessary involution. Trophic victories must be accompanied by atrophic consequences. Thus I force myself each morning to pick up this pen, to joggle my conceptual realities about, to decide on what direction I shall go today. I am both inducing a more effective organ for thinking at the same time that I may be neglecting my strength, my dexterity, my speed and my pleasures by what I do and not do.

Thus make note of the temporal quality involved with human and cellular specialization. If we are forced to perform specific organ or cellular functions over and over again, our excellence and our shape is altered in relation to the time we devote to these different functions. Initially we were molded by an environment that automatically took a cell with nearly infinite evolvement potential, the germ cell, and in the very process of generating greater and greater numbers reached a phase where the non-uniformity of the environment demanded that some functions turn on, while others turn off, so that by the time of embryonal differentiation, three distinct cell masses, with totally different destinies were created. All had common potential initially, yet because of environmental messages, the genetic machinery had to reorganize the governing machinery for the specialty differentiation essential to cellular survival. With continued growth and the greater differential of environments, the specialty of cellular functions became more uni-purpose and most of the individual potentials had involuted in favor of communal potential.

Just as the cell responds quietly to its inter-phasic miracles and changes itself in relation to its stressful messages, so does individual man.

During Para sleep, all the stressful messages of an Ortho day become the electrical and chemical templates that govern the structural growth and recombination essential to depolarizing the electrical template, while at the same time creating a structurally more effective cell or organ for the next day's activity.

We grow and become social creatures because of the outer laws imposed by family, culture, religion, and nation. Outer laws are the resistances to the inner laws that are selfish, egocentric and to a degree feral. Outer law lends value and worth to those items for which we must work and fight. When the organizing forces from without are gone, inner laws gain their hierarchy of pacemaking significance and thus compete selfishly and abusively with each other, creating havoc within the poor human who must bear up under this inner anarchy. It is in this poorly governed human that we find perpetual functional diseases.

In a sense, social law relates to the Ortho dominant responsibilities we assume in the outside world of human relationships. Natural law relates to the Para pacemakers, those egocentric demands we must respond to if we're going to survive. We call them instincts and that is what inner laws are all about. Their demands ultimately take precedence over the demands of outer laws but if we function well in our social world, our private world where Para dominance prevails, will allow us sleep, the assuagement of hunger, thirst and sexual urgencies, and the warmth of a home along with its protection that was not as easily available in centuries gone by.

How can we integrate the electromagnetic science we described initially and the Ortho-Para science we shared, to a small degree, in the latter part of this presentation?

The nocturnal re-creative period, associated with Para dominance, is enhanced by the electronegative or south pole of the magnet. But during the day the electropositive or north pole of the magnet potentiates the creation of the kinetic energy essential for ones day of work and accomplishments. In this sense, the magnetic amplification is that of the dominant autonomic system at work. The circadian rhythms of night and day are controlled by the sun and when the sun rises and we get out of bed the Ortho dominance associated with serotonin, catecholamines and adrenalin and potentiated by the adrenal glands, the thyroid glands and the hypothalamus, are changing the chemistry of the nighttime into the living chemistry of the day. The magnetic polarity of the sun is electropositive. That of the earth is electronegative. At night, when the sun has set and the Para dominance begins to create the changes essential for cellular and organ recreation, we sleep quietly and allow the body to rebuild itself. It is then that the south pole or electronegative side of the magnet potentiates the normal Para polarity of the body.

The Chinese were aware of these polarities and used the term Yin and Yang. When I had an opportunity to compare the qualities of Yin and Yang I found they represented a more ancient awareness of the body's need for balanced opposition to function healthfully and they were primitive representatives of Para and Ortho respectively. Jonathan's treatises deal very completely with the need for systemic and cosmic balance."

Jonathan was very pleased with both Laura and David and quickly made this plain to the rest of the class. He happily said, "I am so happy with the intense efforts of both Laura and David. To this date there are few of the traditional Allopathic physicians who will speak positively about magnetic therapy. This is the case, even though a successful growing industry has developed with devices with variable magnetic configurations that have been used for degenerative and gouty arthritic joints very successfully, and in this way eliminating the use of drugs, all of which have potentially serious side effects which have not been seen with magnets.

Does it work all of the time? No. And this is the case with all therapeutic regimes. But it is a non invasive approach to the treatment of many problems that have no absolute cures and just require remedial therapy to make life more comfortable. It is one of the Alternative therapies that require consideration by all healers."

Chapter 27

Thus far the discussions were probing, and it seemed that Dan had already gotten some good ideas regarding the Center's design, but Jonathan was a little skeptical. He tried not to posture these feelings when he spoke. "It's now time for two non-medical people to approach this subject. Dennis, can an Astrophysicist and an Electrical Engineer add something to the solution of Dan's proble? I've been aware that each couple has been unable to cover each of the subjects that were assigned. Were you faced with the same problem?

Dennis was well prepared. "We've covered the therapies assigned, but I don't think we have enough time to go into all of them in detail and it really isn't necessary. Laser, Neural, Aerobic and Pharmaceutical therapy are all part of Traditional Medicine and discussing them would require an in depth course, which I'm sure is part of the Medical School's curriculum. Therefore I am going to review and try to make some sense out of Homeopathy, a field that many graduate physicians have encompassed, along with many of those therapies that have and will be discussed today.

Homeopathy is a system of therapeutics introduced and postulated by a German, Samuel Hahneman in 1796. Some parts of his system were borrowed from Hippocrates and Paracelsus. The essential tenets of Homeopathy are that the cure of disease is effected by drugs that are capable, themselves, of producing the same symptoms as the disease, if given to a healthy individual. To ascertain the curative virtues of any drug, it must be proved that it causes the same symptoms on perfectly normal individuals. The proven drug is then diluted many times, so that the drug is not detectable by ordinary chemical means. It is this diluted 'provings' that is given to the patient.

Like all not completely comprehensible therapies, Traditional physicians did not accept a theory that was claiming the effectiveness of pure water, since after multiple dilutions there was no medication left and only water was administered.

These objections did not initially prevent the therapy from growing in popularity and there are many good reasons for the patient population to look with some concern at the views of Traditional Medicine. They are both historic and modern. In the Typhus and Cholera epidemics of the 1800s and then again during the flu epidemic in 1917, the Traditional doctors lost more than 50% of their patients while the Homeopathic healers lost as little as 15%.

These statistics are explainable in several ways.

1) The Homeopathic results worked.

2) The Allopathic medicine at that time was lethal in its effect on diseases in which the patient's died in dehydration. The procedures used were a) purging b) blood letting c) emetics and eventually diuretics, the main ones being Mercurial diuretics.

3) The statistics being used were falsified as is the case even in today's occasional frauds.

4) The doctor, his manner, and his method of administering 'provings', both rehydrated the patient, precipitated a placebo effect, and did not administer all the dehydrating procedures used by the traditional doctors at that time.

5) To these, I would add one unknown that I've never read about in terms of the present question, but even as an Astrophysicist I recall reading the differences in the kinetic energy of intracellular fluids, as compared to the extracellular, pericellular fluids. We know that there is a structural difference of water within the cell as compared to that in the extracellular compartment. Even in a non-organic environment, if you add sodium salts to the solution of water, you integrate the water molecules and increase the organizing capacity of the water in relation to the other contents of the solution. In contrast, if you add Potassium to the water, it becomes more disorganized and its kinetic energy is increased. This is shown in the 80mv difference in electrical polarity that makes the cell, filled with potassium, so electronegative as compared to the pericellular fluid filled with sodium. It is this difference that underlies the metabolic and physiologic machinery that creates life. Do the 'provings' that Hahneman created with his dilutional process sufficiently organize the administered water so that it could have an effect on the primary electro-activity of cellular function?"

Cynthia suddenly interrupted Dennis. "Dennis, before you go on, I'd like to throw in a few thoughts on this problem. When we left last month's meeting, I spoke to Cassy for quite a while. She was telling me that one day while she was walking, she wondered what was the best way of examining life in its purest Para form? Could it be compared to the Para of a newborn babe? She suddenly remembered a trip to Yosemite and the reaction she had with her first sight of a Sequoia tree. To her, and then to me, they were the essence of Para.

We were very comfortable equating the polarity and abundance of an unmoving giant to the tiny infant lying motionless on its mother's tummy immediately after birth.

Cassy turned to me with a quizzical look on her face. She said "The essential relationship between plants and man stared me in the face, and yet I cringed, as a physician, when I read that Herbs might influence my healing if I became ill. It is even more ridiculous, since the very drugs I give my patients had their origin in the roots, stems and leaves of plants that were discovered in the fields and forests all over the world by Medicine men, Shamans and Witch doctors who were keenly aware of what botanicals successfully influenced the health of the local tribes. It was with the knowledge of these Shamans that the Pharmaceutical Industry put their science to work and successfully extracted the chemicals of worth and then synthesized them, using the modern techniques of chemistry today. With this in mind, I have had to open up and read with greater acceptance that there is more to herbal medicine than my doctor's mentality is willing to accept. In thinking on this problem, I realize I had put myself into a dilemma when I recognized that the pharmaceutical industry controlled the puppet strings that were attached to my pen when, each day, I was writing prescriptions for my patients over the last ten years. Yet there is the other side of the story. Herbal medicines do not always have what they claim they have in them. There are stories of herbal contamination which make sense to me so that along with new knowledge of their possible therapeutic worth I find myself concerned over the dangers associated with using them. Cassy owned and verbalized the existence of a serious problem that my own personal physician also speaks about. With this in mind I bought a new book written by Rudolph Ballantine M.D. entitled 'Radical Healing'.

He is an M.D. Psychiatrist who attempted to Integrate Medicine. It surprised me that with his traditional training, his writings were not too kind to Traditional Medicine and he radiated with confidence the benefits of all Alternative approaches to the maintenance of health and the healing of human illnesses.

In reading Ballantine's book I found many scientific errors and concepts that rested on a pseudoscience designed to confuse. He had suddenly become a Homeopathic doctor. In reading his multi-dilutional concepts, outside of the introduction of new words, with logic that defied reason, he redefined the relationship between molecules and biology and more than that, he gave me a major headache.

Now that I've criticized the poor doctor I want to step back a little. He speaks of the mechanisms that Candice Pert spoke so much about in her book, 'Molecules and Emotions'. He spoke of the mind-body powers that positive assertions could reinforce positively, such as the psycho-neuro immune defenses, and in contrast the negative assertions that could do the opposite.

Ballantine then talked about patient types and each of their vulnerabilities. As an M.D. and Psychiatrist he was now talking in his element and to a degree he was agreeing with some of the concepts of Jonathan. He made a great deal of sense, especially in this age of a pleasure oriented population that appears to be self intoxicating during youth, with poor health habits that are family intoxicating, socially intoxicating due to relational pathology and these are superimposed on those due to industrial wastes and air pollution, so that because of this, we are not as capable of handling the viral and bacterial population that shares and has always shared this Earth with us. Historically the plagues struck the world population when the civilization had morally sank to a depressive low and the individual and societal immune systems were concomitantly depressed. When enough people had died, the plagues ceased even though the vector and the cause never really completely disappeared. Without apparent reason, it was time to bury the dead and reevaluate our individual and systemic lives and find out how to alter our behavior, eating habits and intimate relationships. Certainly, out of every pandemic came a new wisdom, new inventions and a change in our world that became the foundation for new growth which peaked, just before new evidences of Individual and Societal deterioration again began to appear.

Ballantine focuses properly on the faults associated with Partistic Medicine and offers an Alternative which has several properties I admire.

1) It turns the Rescuer-Victim relationship into a teacher-student relationship.
2) It demands a careful search into the whole and therefore a detailed history so that not a part, but the entire patient can be treated.
3) The very nature of the time dedication, focusing on the intimate relationships, is in this way establishing a powerful therapeutic effect on both the healer and the one to be healed.
4) Suddenly the "Whole" is being treated instead of a "Part" and the power of an awareness of constitutional polarities are being included in the advice and prescription being given. In this sense the Ortho-Para polarity and its significance in disease evaluation becomes relevant to that which is being treated, and the patient's symptoms, rather than a specific diagnosis, are being treated.

This M.D.-Homeopath, whose practical observations demonstrated his interest in a holographic interpretation of different parts of the body, revealed maps of the face, hands, tongue and wrists that were not different from those we see on the soles of the feet when studying Reflexology.

His all-around competences and his frequent changing of treatments was suggestive that he himself was still searching. He changes his treatment from vitamins, minerals, herbs and diluted Homeopathic nostrums so quickly that one wonders whether he is giving enough time for the patients to get well by themselves. He claims that he himself is his most difficult patient. I truly believe that he is brilliant, but misguided. His belief in the intuitive mind has given him more symptoms and more miracle cures, and his extensive studies have been in fields that are strongly intuitive, imaginative and creative, all requiring a minimal of scientific proof. He would claim that reductionistic, partistic thinking can only fall short of absolute truths and with this I agree. But it doesn't therefore open the door to any impulsive, intuitive madness that happens to fly your way. With this having been said, I close the book on Ballantine and turn the floor back to Dennis."

"Cynthia, you said a mouthful. It is apparent that we haven't discarded Homeopathy completely and there are times that you have to go with the therapeutic evidence even when scientifically it makes no real sense. We gave several reasons why the therapeutic atmosphere surrounding the teacher-student relationship might itself be the remedy for many illnesses. The one scientific thought related to 'provings' is the distant possibility that it may organize body fluids due to non measureable factors that have to do with integrating peri-cellular fluids. I'll leave that to the medical profession. As an Astrophysicist I believe this will be my last attempt at trying to understand what doctors do."

Jonathan took the floor with a huge smile on his face. His Astrophysicist and Electrical Engineer students did their homework very well and he was pleased with their dissertation. Dan raised his hand and Jonathan recognized him.

"Can you tell me what facilities this Homeopathic representative would need at the Center?"

Dennis replied "From what I read in the Homeopathic, Chinese and Ayurvedic literature, you will need a comprehensive herbal pharmacy with a botanically trained pharmacist. The size of this pharmacy would depend very much on how complete the herbal compendium will be. I think the one to ask would be Dr. Su Shi the next time you meet."

Chapter 28

Jonathan looked at the class with a big smile. "I must admit that the in-depth studies that preceded this day of sharing have been a lesson to me and I guess to everyone in this room. Steve and Ariana, I guess it's your turn to add to our knowledge".

Steve stood up and faced the group. "I don't believe we were assigned the most difficult Alternative therapies. For that matter I never conceived that some of ours were really classified in the Medical armamentarium, because they are, in most cases, a significant part of everyone's life. I'm talking today about a few of the obvious subjects and Ariana will stop me when she determines, and she always does, that I'm talking too much."

David chimed in "How long did she take to figure that one out?" The class broke up in laughter.

"David, she's been aware of it for the same 25 years she has observed that your mouth is usually chattering when you should be listening." Another response from the class silenced Steve for a moment as he stood smiling at his classmates.

Steven then continued, "I'll start my talk by alerting you to the fact that the living process gives us all a balanced share of Pathos and Bathos. The symptoms of Pathos are sadness and depression and could be classified by Jonathan as a Para emotion. To balance this extreme, we are blessed with the Bathos, the humors and occasionally the ridiculous side of life. Bathos, in the form of humor, litters the Email of every computer geek and adds a laugh, giggle and occasionally the feeling of wonderment when the positive tickle of life allows us to be positive, uplifted and Ortho dominant in our emotional posture. If you examine Jonathan's first Ortho-Para Treatise, on page 526 and 527 you will see the autonomic unfolding processes of an infant during the first two years of life. Delight and elation evolve during the first year, while sorrow and grief can occur so quickly that an infant who loses its mother within the second 6 months of life, can actually die in anaclitic depression.

Since sadness and happiness, when not extreme, are the emotional equivalent of either a negative or positive Para-Ortho polarity and fear and delight, when not extreme, are the emotional equivalent of either a negative or positive Ortho-Para polarity, both are frequent emotional polarities in a perfectly well balanced life and have a powerful effect on human behavior in every system of life. Humor tends to precipitate the positive polarities of life and is in contrast to tragedy, which more easily precipitates the negative polarities. Their behavioral responses are directly related to the emotions evoked by what we hear or see.

Every perceptual organ, our ears, nose, eyes, tongue and the sensory perceptions of our skin are powerfully capable of moving the emotional pendulum toward one or other extreme and it is humor's goal to move it to a positive polarity, whether it be a positive Ortho or Para.

Arthur Koestler, in his masterpiece 'The Act of Creation' spends a great deal of time defining the many types of humor. He says, "Humor is the only domain of creative activity where a stimulus, on a high level of complexity, produces a massive and sharply defined response on the level of a physiological reflex. The reflex is usually in the form of laughter."

However, in examining the many tripping mechanisms of laughter, there is evidence that Bathos can easily turn into Pathos without the logical evaluation of its causes. I think that its use as a therapeutic agent has to be seriously evaluated and then understood in relation to the person and the illness which is being treated.

We must remember that just as talking is a means of relieving tension, laughter is an energetic way of discharging tension but it is frequently preceded by a story that has added a bit of tension of its own. Talking and laughter are behaviors directed toward the outside world and as all other behaviors regarding the outside world, there is an Orthosympathetic (Ortho) component that represents the response to and release of tension. The concept is not new. The first to make the suggestion that laughter was a discharge mechanism for nervous energy seems to have been Herbert Spencer.

Let me quote a paragraph of Spencer. "Nervous tension always tends to beget muscular motion and when it rises to a certain intensity…the emotions and sensations tend to general bodily movement. The movements are violent in proportion, as the emotions and sensations are intense". He concludes "When consciousness is, without awareness, transferred from great things to small, the liberated nerve force will expend itself along the channels of least resistance which are the muscular movements of laughter."

Obviously laughter is not the only form of tension release, but it is one that is a reflection of aggression or apprehension, without purpose or goals, and merely a discharge of un-channeled energy. Physiologically, as described by Koestler, as we listen in on any conversation, we are receiving an energy charge. It is merely the opposite of the discharge that takes place when we speak. While listening, our upper brain is charged with information from the outside, at the same time that our hypothalamus and amygdala of the midbrain are controlling the hormonal discharges from the endocrine glands. If there is a sense of outside aggression, the neuro-hormonal discharge is of Ortho or adrenergic polarity. If there is a sense of sympathy or compassion, the neuro-hormonal response is Para incorporative.

Laughter is a discharge of the emotional or adreno-hormonal response, when the upper brain suddenly realizes that the logic of the comment makes no sense and is not the dangerous aggression it had anticipated.

Crying, on the other hand, is a discharge of the emotional or Parasympathetic hormonal responses. It is when the upper brain realizes the tragic consequences of a comment, accident or death.

Koestler brilliantly described the anatomic behavioral and temporal components of both laughter and crying (page 272) so that we see the pendulous swing that occurs as we experience the joys and sadnesses of life and the emergency mechanisms that come into play when we become polarized too intensely in either an Ortho or Para extreme. But we must note that the pendulum swings all of the time and that the intensity of a particular emotional or autonomic dominance is in proportion to the width of the swing and whether the swing returns to a median point or dangerously swings to the opposite extreme, thus producing a whole new set of emotional consequences.

For this reason and for the fact that humor and laughter can burden the already autonomic extremes that exist with illness, I would think that careful choosing of the type of humor, in relation to the emotional polarity of the patient, is in order. I would not encourage an enormous Ortho response in a patient who is in cardiac decompensation or a Para response in a patient who is receiving chemotherapy for a malignancy. Keeping the pendulum as close to its median position, especially when dealing with the ill, would seem a therapeutic precaution that is necessary." Steve hesitated for a moment and Ariana stood up and faced the class.

"I think Steve has pretty well established a fact that probably is true for all therapy regarding the ill but is even more important because it lies at the foundation of good health and Wellness. There is no one treatment that can be used on everyone and anyone. An in depth awareness of the Ortho-Para polarity of each patient and the Ortho-Para effect on the life of the healthy and the ill should remain part of the clinical acumen of every doctor or healer involved in the care of a student of life, the patient.

In contrast to Steve I've chosen a few Alternative therapies that have already been touched on. Light physiology was somewhat described when Laura and

David discussed magnetic therapy. The Pineal gland and its serotonin and melatonin output magically influences our nyctohemeral (night and day) rhythms and when these rhythms are disturbed, whether due to too much light or an inadequate amount of light, it can influence our autonomic polarity, our sleeping rhythms, and the behavioral decisions made by those who are seriously symptomatic because of a pathologic depression, with occasional thoughts of suicide.

If you check the suicide rates in the nations that inhabit the polar regions of the earth, you will find it highest in Finland, Sweden and Norway and the incidence of alcoholism and depression parallels the rates of suicides.

We apparently need a certain amount of exposure to light and the northern magnetic field, and the more Para we are in our polarity, the more vulnerable we are to the depths of depression that precipitate feelings of self derision and self destruction.

In searching for the pacemakers of body rhythms, the organ that stands out as a modifier and potentially controller of circadian rhythms is the Pineal gland. There is little question that this nut size gland in the brain is the light sensitive organ that controls numerous bio-rhythms. We all experience mood changes that are related to dismal, cloudy days and bright, sunny days, but there are those who suffer with a condition known as Seasonal Affective Disorders (SAD) and fear the onset of winter, even in the temperate zones, but have far more dread of the short days of the polar regions when their symptoms of depression intensify dramatically.

There have been many theories that have been offered regarding a healthy life and longevity. Have you ever wondered what changes a soft, innocent, guileless, happy, laughing infant into a hardened, guilt ridden, somber, wrinkled and sometimes wisened senior?

It begins immediately but is unnoticeable for a long time. Under the influence of the sun, its radiances and its electropositive magnetic field, we awaken each morning with a surge of energy and begin directed movements toward our goals. The behavioral response to these goals redesigns our structure and eventually carves us into what the past has helped us become.

When the sun sets and night comes upon us, our energy slowly dwindles. We are encompassed by the electronegative Earth's magnetic field and with it we eventually sleep, the process that is necessary to recreate, regenerate and redesign our bodies in response to the resistances and stresses we faced the day before. Dependent on the goals, we are stronger, faster, more skillful and at times smarter and more intuitive than we were, and we're far more comfortable with our daily performances if they are similar to what we did the day before.

If however there is an inadequate amount of sleep and therefore a decrease in the time of the healing associated with sleep, we have altered the create/destroy ratio and we've lost a part of ourselves- a minimal or maximum number of cells at a time. This is the onset of the deteriorative changes that take place when we are destroying structure during the day and don't have enough time at night to recreate it.

The controlling organ that paces this circadian rhythm of night and day is the Pineal gland. In the daytime it secretes serotonin that enhances the energy creating mechanisms. It also plays a role in stimulating the secretions that enhance the adrenal cortex and the Orthosympathetic Nervous System. At night, when the photic stimulation of the retina is gone, serotonin is changed to melatonin and the restructuring hormones and the Parasympathetic Nervous System take over.

When we're young, there is a huge amount of melatonin present during the night, so that the creative mechanisms dominate, as can be seen during the growth phase of every person. But as we age the melatonin gradually decreases along with sleep time and our ability to regenerate the architecture of strength and agility is handicapped. Thus, with age, we end up going to physicians and physical therapists with the incapacitating diseases that limit function, cause great discomfort and eventually are associated with permanent disabilities.

From what we now understand, during the day the amino acid tryptophane is turned into Serotonin by the Pineal gland. It is secreted into brain tissues as a neuroendocrine and participates in facilitating the workings of the dendritic circuits of the brain. It is an awakening neuroendocrine that is essential to the production and sustaining of kinetic energy. When low or blocked, there is a tendency toward fatigue and if long lasting, depression.

When light hits the retina, it is transduced into an electric current that passes along the retino-thalamic tract, behind the eye, to the suprachiasmic nuclei of the hypothalamus. The stimulus then continues, via the Pineal nerve (an Orthosympathetic nerve) to the Pineal gland. The above reaction forms serotonin, an Ortho endocrine and blocks the chemical manufacturing of Melatonin. However, when light is removed, so is the stimulus via the suprachiasmic nuclei and it releases the inhibition of the formation of an enzyme N acetyltransferase that is able to turn serotonin into melatonin and in this way alters the biologic clock so that nocturnal Para activities can assume its functions. The same stimulation to the formation of Melatonin can be accomplished with acetyl choline, a Para hormone, so that an increase in Para or a decrease in Ortho can accomplish similar changes in mood and autonomic polarity.

We therefore assume that serotonin precipitates an Ortho stimulus and melatonin its Para antithesis.

In the above sense it appears that the Pineal gland is photosensitive and can understandably be called the third eye that paces and oversees many of the circadian rhythms of the body. Every endocrine organ has its own intrinsic rhythm that is, to a degree, controlled by a higher rhythm from above. The higher rhythm from the brain is a more sensitive pacemaker to which the intrinsic organ pacers are able to respond. It is therefore both the more primitive Endocrine system and finally the Autonomic Nervous System that adjusts the physiologic rhythms so that they are more responsive to the activities of life.

With age, there is a progressive calcification of the Pineal gland with a slow decrease in Melatonin formation and though there is no definitive evidence, there are many who believe that aging could be slowed or inhibited if adequate function of the Pineal gland could be maintained. There are those who are giving supplemental melatonin to accomplish this end.

The following are a small list of medical claims being made regarding the Pineal.

1) Too little melatonin is one of the causes of aging.

2) Jet lag is due to a decrease in melatonin due to bio-clock abnormality.

3) Seasonal Affective Disorders and Bipolar Affective disorders are due to too much melatonin and not enough light. The serotonin levels are low and drugs that enhance serotonin action also help eradicate the depression.

4) Melatonin coordinates Fertility.

5) Melatonin gives a restful sleep and therefore enhances all of the Para functions of sleep.

6) Melatonin is found to be deficient in some cancers which are now showing a response to Melatonin administration.

7) Melatonin has a stimulating effect on the Immune system via its effect on T lymphocytes.

8) Blind women have more melatonin and a more powerful Immune system and less cancer.

I realize that this is a mouthful, and sounds more like a pharmaceutical advertisement, but in line with the physiology already described, all of these claims are possibilities and certainly should not be completely ignored.

In the human, the peak secretion of melatonin is about 3 A.M. to 4 A.M. The complex relationship between the Pineal hormones, the other Neuroendocrines, the Endocrine hormones, the Autonomic nervous system and the Central nervous system is still a huge area for research and I'm sure it's going on today.

Electromagnetic waves of other frequencies are used and are essential to normal function. The ultraviolet wave lengths are essential to the production of Vitamin D. The infrared wave lengths are used for deep heat therapy. X-rays are used for diagnostic procedures and also, in higher frequencies, for the destruction of cancerous growth. Gamma waves are used for the treatment of cancer. Radio-waves enhance the magnetic fields of MRI's and are used for communication. I am sure that I left out many of the wonderful uses of low and high frequency electromagnetic waves, but I'm going to stop now.

The more I studied the laws in Jonathan's first Treatise, the one thing I found certain is that the Para/Ortho ratio is highest at birth. It controls the create/destroy ratio, the anabolic/catabolic ratio and is effected by many levels of biologic and physical function. We assume that the Central Nervous System can control, to a degree, the Autonomic Nervous System which controls feelings. In turn the Autonomic Nervous System can rise above Neuroendocrine controls, which control all hormonal secretions. It is also claimed that Neuroendocrine controls can rise above the Endocrine controls which are the most primitive systems of controls that are attached to human phylogeny, but studies always clarify the feedback and the biocybernetic mechanisms that create the Neuroendocrine axes that ultimately seek balance in relation to our goals, rituals and systemic responsibilities."

Chapter 29

Jonathan stood up and looked at Ariana in amazement. "I had no idea that you'd be able to research your subject so thoroughly, being that you're a lawyer. It frightens me that you lawyers are mastering the biologic sciences so well. I understand now why the insurance premiums have gone up so high."

The class interrupted Jonathan with laughter. He continued, "All of you have gone into your subjects so thoroughly I can see why you've been able to cover only a few of the assigned Alternative Cares." He turned to Kenneth and Allison.

"I realize that we still have not heard from Beth and Jason, but I want Kenneth and Allison to first help me, once again, to reconstruct the basics of life, its stratified structure and function and try to build a normal human with all of his complexities and having done that, to try once again to analyze the causes of Illness in relation to these structures. In that way, we can more easily determine which of the dynamics of Traditional Medicine or any of the present Alternatives, might be more favorable and a less expensive route for achieving Health and comfort again.

If there is any one of you who wants to join this construction crew, you are welcome to join in this task of building Man out of the stuff so generously given to us by the forces of Cosmic Creation. Twenty five years ago, when you were ten, we created 92 elements out of Mattergy. It was you who defined Mattergy as a form of undifferentiated Kinetic Energy that followed the fifty nine laws of Systems and Particles. These Natural Laws automatically influenced the particles of that day and after fifteen billion years it still continues to influence the behavior of all of life's creations, including the life and death of all Mankind.

However I realize that before we begin a dialogue with Kenneth and Allison, I believe Steve is in a better position to talk of that moment when something magical took place, and out of clay in the ocean bed, the first life form was created."

Steve stood up and rapidly responded to Jonathan. He was thrilled with the privilege of being able to discuss the creation of life from innate, inanimate matter. He was radiating this confidence when he took the floor.

"Thank you Jonathan. I'm fully aware of the privilege I've been given to give a logical account of how it all began so many years ago. Thus far we have no evidence of life anywhere else in the Universe, although we theorize that there are other planets, in other galaxies which have similar geologic, geodesic and atmospheric conditions to ours, but there is yet no absolute evidence.

We must use our imagination along with the evidences that have been presented to us. Surrounded with clouds of Cosmic and Terrestrial dust, the hardened crust of the dark earth was being invaded by Gamma rays of Cosmic origin. The diastrophic terrestrial volcanism recurrently lay open the magmacious cauldron below, at the same time that mountains rose and the sea beds sank, excavated by fiery lava flows. When it was done, the earth had its one and 1/2 billionth birthday. Its mountains were quieter as they rose high above the sea beds and what were deep valleys became the oceans, seas and lakes of the world.

It's not as if all was suddenly quiet and still. The young earth was a child of the Cosmos. Its birth was part of a solar birth, secondary to the cataclysmic death of a secondary star. The very star dust that merged eventually contracted and formed our sun and all of its planets. It was part of the Cosmic experiment that began nearly twenty eight billions years before, so that what appeared to be chaotic disorder was truly part of the grand scheme of Cosmic evolvement.

The atmosphere was hot and the water was in constant transfer from the sea to the heavens, a process that continues on earth to this day. With each storm the rains eventually produced mountain streams, rivulets and rivers that washed the elemental debris back into the seas where it remained and slowly concentrated over a period of time. It was therefore in a sea, filled with the elements created in the Galaxies and the Primary and Secondary Stars that this primitive solvent, exposed to heat, electrical storms and a chaos of electromagnetic radiations of countless frequencies that the chaos and kinetic energy of the sea required a molecular resolution to this crisis.

There are two classifications of science dedicated to the answering of the questions as to how this resolution came about. The first, who we now call Constructionists, were committed to gathering the elemental substrates of life and creating life out of the same material that is the theoretic inorganic soup that pre-existed the creation of life.

The second, the Reductionists, worked with ancient forms of life and organic fossils and were the scientists who had already dated the beginning of life to be three and one half billion years and who were seeking older and older material for evidence relating to the first fossils and their chemical composition. Thus the analytic Reductionists were working back from the complex to the simple and the creative Constructionists were working forward, from the most elemental and simple to the most mega-molecular and complex.

Of the original 92 elements however, the Constructionists, our modern alchemists, recognized that the four most abundant bio-elements were Hydrogen, Carbon, Oxygen and Nitrogen. Organic chemistry is not the same as physiological chemistry in that in the former the solvents used for discovery are many. In the latter, the only solvent essential to life is water. Thus, the direction of study for the Constructionists was to see what large molecules could be created out of basic elemental materials in the theoretic environment of early earth. They successfully created, via these techniques, a large number of bio-molecules that were successfully made into what they called Proteinoids.

Based on Darwin's view on natural selection, it was in 1865 that a germ biologist by the name of Ernst Haechel gave lectures on the origins of the first cells and it was he who offered the first theories regarding the processes of self organization that is presumed to be the cause of the first created form of cellular life.

It was not until 1924 that Oparin reiterated the views of Haekel but with greater insight into its possibilities because of some of the new advances made in science. But it took another thirty years before their ideas on the spontaneous formation of biologically significant molecules from elemental precursors could be verified in a laboratory. In 1955 a science known as Chemical Evolution surfaced. It dealt with the abiotic and prebiotic origins of primitive small molecules of organic materials. In time it was referred to as Molecular Biology.

Under slightly lower pressures than the four primary elements of life, the Carbon, Hydrogen, Oxygen and Nitrogen already mentioned, the cations, Sodium, Magnesium, Potassium, and Calcium plus the anions Carbon, Phosphorus, Sulfur, and Iodine were created. It was these eight latter elements that controlled the functional elements of life. It was in contrast to the first four elements that created most of the structural aspects of life. The total number of atomic elements that make up life are twenty five and are represented by the first twenty five elements of the periodic table, excluding the Noble elements. But Jonathan, now that I went through this long tale, and that is what it is, what has it to do with Medicine?"

Kenneth immediately picked up on the question. "It would be an excellent way of beginning the study of Medicine and the physiology and chemistry of what we could call 'The Layering of Man.'" Laura, with bubbling enthusiasm, interrupted, "What you're saying is that if you know the structural and functional layers of Man, you've not only learned what you must know about Man, you will have touched on the Ontogenetic steps of Man's Phylogenetic creation. This, in turn, would help us to understand more completely the entire evolving process. I knew ahead of time that this subject would be discussed, so that to aid our studies I wrote an outline of the evolving process that should give direction to our focus.

It was not surprising to learn that life began and grew in a Hydrogen saturated atmospheric environment. It was a reminder that the anabolic processes of regeneration and reconstruction in living organisms is a Para reductive, alkaline process, using as its substrate the Ortho, catabolic, acid producing, degenerative bi-products of oxidation that take place during the day when potential energy is being changed to kinetic energy, structure into function, and all for the purpose of normal daily living. This would suggest that even prior to life, a Para chemistry was needed to incorporate the chaotic pre-biotic environment. As has already been suggested on several occasions, the Para process was the magic designed to incorporate the evidences of the oxidizing, acid producing chaos of the Ortho processes of catabolism and Living."

Jonathan quickly interrupted, "Laura, that's wonderful. I had heard that Oparin and Haechel and later Harold Urey postulated the 'Reducing Atmosphere Hypothesis' but I never related it to the Para reductive anabolism of the modern cell. Your contention is another verification that life began in a reductive environment and all the evidence we now have suggests that life, up to the beginning of the Annelida, had only a Parasympathetic Nervous System and that the Chromaffin precursors of the Orthosympathetic Nervous System did not arise until intrinsically energized movement began.

It might also be suggested that the photosynthetic Fauna probably came into prominence some time prior to the Annelida so that the functional mobilization of animal structure, controlled by the Ortho polarity, would have arisen at that time. The processes of contraction and expansion, which require a Para reductive, reconstruction polarity and an Ortho energy producing source for mobility, came into existence when mobility first began."

Laura listened to Jonathan and gleamed joy at his comments. She went over to him, as she had done many times previously when he responded with approval to what she added to all conversations and embraced and kissed him.

Kenneth laughingly commented, "Laura, you haven't changed in the last twenty five years. You still are brilliant and know how to get on center stage. A long time ago Allison told me to watch you at work and she predicted that you'd be starting the first Ortho-Para Professorship at a major renowned University.

But I'd now like to respond to Jonathan's request of Allison and myself. Obviously the first Layer of Man is the first 92 elements, only 25 of which are now known to be part of Man's elemental history. I would not be surprised if one day we discover that the remainder are also essential in trace amounts. From these Monomers, in the hands of Urey, Fox and the students in the Constructionist family, polymers were created that are essential to life. However, the creation of Life itself has been elusive. I'll comment on that failure when I conclude.

The Constructionists may not have created life but they have surely created most of the molecules, including the 20 essential amino acids, but the complete long change nucleic acids, DNA and RNA have yet to be produced experimentally. There may be a very important reason why we've failed in our search and that too will be commented on when we talk about why the discovery of a process to create Life will not happen.

The second layer is the binomers, such as H_2O, CO_2, N_2 and a myriad of two or three element polymers that are essential to life. When you consider that physiological chemistry grows as a process of the increasing number of elements, we can claim, on a polymer level, that there are thousands of layers that are all essential to life. Add or subtract a monomer from the polymer and we may have a different disease.

Beyond the elemental layers we face the complexity of the cell itself and the many organelles that make up the cellular community. No different than a city and its people and industry, the organelles play life sustaining functions that ultimately may influence the life expectancy of the cell itself and the human in whom it resides. In fact, the laws of Systems and Particles really come into play as we list the layers and find out that each is a system made up of particles that are responding to the laws listed in the first treatise.

Thus, from the first element Hydrogen, we have seen the process of reduction producing all 92 elements.

From the birth of all the elements, we have seen the process of reduction creating all the monomers and polymers that make up the chemical porridge that remains the basis for Biochemistry and the Physiologic processes of metabolism.

With the birth of the cell, we suddenly discover a city of organelles, each a biochemical city unto itself, with a distinctive chemistry essential to the function of the incorporating cell and to the super hierarchy encompassing it, the human. Each year we find new functions within the cytoplasm, the nucleoplasm and the genetic community of assorted chromosomes, which are also merely a community of intelligent, genetic architectural particles that govern the production and reproduction of working proteins that not only become the substructure of the cell, but direct the community of cells that make up each organ of the body.

Damage to any portion of the cell that oversees a vital function is compensated for as long as there are enough cells with similar functions, so that the number of cells performing similar functions act as a safeguard for the hierarchy that depends on and rests on an adequate number of these specific and specialty oriented cells. When the population number falls below a critical number of functional cells, we have Illness.

We can look at the various organelles in the same way. If the number of normal mitochondria within a cell population falls below the level needed for energizing cellular functional competence, we have Illness. No less than a factory, the output of any product is dependent on the competence of the workers (the organelles), each of whom is comparably represented and enclosed within a living semi-permeable cell membrane

We see a pattern to this biological urban growth. The organ is a city of multi-purpose cells. An organ system is a city of multi-purpose organs and although the complexity of natural biological growth is magnified when we encompass all organ systems into one human being, it is only the beginning of a new system based on social rather than the natural laws that preceded it.

The complexities of social growth have been described, in some detail, in the three treatises and Novels that Jonathan has written. Thus, in the birth of a baby a new hierarchy begins. This time it is not of the elements, monomers or the polymers of biochemistry, or of the cells, organelles, organs, or organ systems that become the stepping stones to growth and maturation. Each step up the ladder, from element to infant, to the complex social systems in which people work, may be the eventual source of Wellness and Illness. It is our job to discover the etiologic site of malfunction, up or down the hierarchy and whether it is in the elemental base of life or in any of the cells or organs within which human functions are acted out.

The movement up the ladder of Evolution started with Monogalactic chaos and slowly, step by step, the undifferentiated chaos was incorporated into the protons and electrons that became the base structure of the 92 elements of the Periodic table. From the beginning, there has been one commonality in the building (or birthing) process and that is the use of Hydrogen (a proton) as the primary substance or fuel of all elemental creation.

For Oxygen to become water, it must be reduced by Hydrogen. The entire building process of the Organic, Biochemical, and the Physiological processes of metabolism rest on the fusive forces of Hydrogen and its sister elements. Every bit of the Potential Energy, the structures of life, rests on the cohesive talents of the subatomic and atomic contents of each cell.

It therefore came as no surprise when I realized that the first structure, and the base of all structures that followed was the first element, Hydrogen.

We note that the evolvement process of matter, life, man, the environment, and the civilization in which it took place, was first filled with Chaos of varying intensities. The first creation of neutrons took place in a cauldron with a temperature that was close to ten thousand, nine hundred thousand million degrees.

Diagram G (from Treatise I)

Threshhold Temp-(degrees)	Particle	Bond Strength
10,900 thousand million	Neutron	939.55 MEV
10,888 thousand million	Proton	938.26 MEV
1,566 thousand million	Pi-meson	134.26 MEV
1,226.2 thousand million	Muon	105.66 MEV
5.93 thousand million	Electron	.511 MEV

Excerpted from Weinberg's "The First Three Minutes"

(What follows is the author's imaginative extension of the above.)

First Period

30,000 million	Hydrogen nuclei	939.55 MEV
1,000 million	Helium nuclei	27.53 MEV
300 million	Hydrogen and Helium	

All Nuclear processes have stopped.

Second Period

300 million	Lithium > Fluorine

Third Period

200 million	Sodium > Chlorine

Fourth, Fifth, Sixth, and Seventh Periods

100 million	Potassium > Bromine
	Rubidium > Iodine
	Cesium > Astatine

Eighth Period

15 million	Francium > Lutetium

The Sun

By the time our Sun was created, many Multi-galaxies and Multi-star periods of creative chaos had already taken place and the temperature was only 15 million degrees. Matter, the subatomic particles and the elements had already been created and countless cycles of expansion and contraction had already taken place.

As the heat of the crust of the Earth subsided, there was still a pressured contractile focus, within the Earth's center, of gravitational origin, which kept the Earth volatile and was the cause of the ultimate creation of mountains, rock and clays and the seas and lakes, so that the next steps in creation would still require heat and a stress (a lesser chaos) due again to systemic gravitational contraction and the bonding of those elements that could light the fires of life and start the next phase of evolvement on the soon to be living Earth.

Remembering that the fires of Creation are due to contractile exothermic reactions, if we examine the qualities of contraction we note the following:

1) Increased Systemic density
2) Increased Systemic temperature
3) Exothermy
4) Decrease in volume
5) Increase in pressure
6) Increase in Kinetic Energy/unit space
7) The contractile systems are filled with expanding particles
8) Inherent in a contracting system is the potential for uni-polar creations
9) Time tends to contract (speed up)
10) When creative, it is non-entropic
11) Creates antitheses
12) It is basis for reduction and human anabolism
13) Spirit, bonding, and love are manifestations of contraction.

In contrast, Expansion has the opposite qualities:
1) Systemic densities decrease
2) There is decrease in temperature
3) It is Endothermic
4) ystems increase in volume
5) Decrease in pressure
6) Decrease in Kinetic energy per unit volume
7) Expanding systems are filled with contracting particles
8) Inherent in expanding systems is the potential for bi-polar creations
9) Time tends to expand (slow down)
10) When creative, it is entropic
11) Destroys or integrates antitheses
12) Is basis for oxidation and human catabolism
13) War, conflict and hate can be manifestations of expansion.

The qualities of contraction have the identical properties of the Parasympathetic Nervous System.

The qualities of expansion are identical to those controlled by the Orthosympathetic Nervous system.

Whereas Ortho becomes the source of outer chaos due to our interactions with the outside world, Para becomes the encompassing Para contraction that ends in a relieving explosion and a creative implosion. The implosive manifestation of Para are seen in changed inception (feelings), changed perceptions and as we age, changed conceptions. Human growth and the weight of influence of the contractile creative power shows its influence on human behavior in the way it slowly influences our evolvement. In Human terms we call it maturation.

That first moment of life that the Constructionists have unsuccessfully sought was as much a part of the creative cycle as the events we've already noted. For that matter, it was the biologic contractile chaos that preceded the creation of the initial cell that triggered the Para polarity that encompassed the pre-biotic chaos, and which then created the explosion that relieved the mounting contraction and the implosive contraction that became the first life forms. It is quite possible that it is not only the explosive reactions that are necessary for any creative process to take place but they must be associated with an implosion that is capable of repackaging the reactants into a different level of evolvement or maturation.(See the laws of Systems and Particles in Ortho-Para Treatise I).

Law # 22 states 'One cannot measure the qualities of a contracting system.' Because of that, any contracting system is not available for study via the routes of Science and Mathematics. Man has consistently failed to understand and measure the processes of creation, whether it is matter, the Universe, life, the thoughts of man, or how the pre-creative chaos or stress that precedes every major structural change comes into existence in the brain, and then influences the outer behavioral expressions of the changed structure.

It is obvious that our direction of maturation will be influenced by the types and qualities of daily Ortho stress we ultimately incorporate during the nocturnal Para dominance of sleep. It is during the stressful polarity, associated with the goal oriented activities of living, that we create the chaos (or struggle) that Para sleep incorporates into the structural changes, and that when repetitive, become the thoughts, feelings, powers, behavior and the physical appearance and competence of tomorrow. If we don't act out each day with physical and mental activities appropriate to life, we accept a prolonged Para dominance which leads to Para diseases. If we act out excessively to life's demands and don't have an adequate period of rest in Para dominance, we fall prey to Ortho destructive diseases.

The list of Ortho and Para diseases are immense and the very causes of the Illnesses are too frequently an imbalance in the cellular Create/Destroy or Para/Ortho ratios. It should therefore be apparent that too much cellular Para creative time will cause cellular dysfunction and cancer, and too much cellular Ortho destructive time will cause Cardio-vascular diseases and organ breakdown.

If we learn to understand the body, its strengths and weaknesses, we can design goals that will enhance or depress our potentials for the future and by classifying our physical and mental status at any moment in time, we should be able to design a therapeutic schedule that would change the Para/Ortho ratio to a more favorable one for improving our health and well being.

Each layer of our multi-layered structure can be the reason for good health or for a destructive or an over-productive abnormality that can be lethal. Although the diseases of Ortho expansion have been successfully treated by present day Allopathic physicians, those of Para Contraction, the cancers and the auto-immune diseases, resist our understanding and are very much more difficult to treat.

Understandably, in today's field of Medicine, the Para diseases are treated with destructive and toxic drugs and limited expressions of Ortho polarity in order to counteract the pathologic Para disease processes. The Ortho diseases are treated with palliative medications and time, which allows the Para polarity to hasten the healing time.

It is here that Alternative Care Therapy is capable of producing the Para intensity and the prolonged time essential to treating the degenerative diseases of Mankind."

Chapter 30

Jonathan realized that it was Helen's and his responsibility to incorporate all of the studies that his 17, plus Si Shu and Shang, had expressed during these hours of exploration into the Alternative fields of health care. He was aware that Shang and Dan had already devised a system for organizing the variegated therapeutic techniques, so that they fit into three different classifications of therapy, but intuitively he felt it would not be enough and that he and Helen would have to do a better job, more difficult to refute. In a moment when they were alone, he would share his concerns with Helen.

Shortly later the opportunity presented itself and he spoke. "Following the wonderful presentations of the class, it is our responsibility to present some sort of integration, so that the concepts of Integrated Medicine are justified by a more encompassing paradigm that both Medicine and the many Alternative procedures that have successfully been used throughout the world, can find justification, because with this in mind, it would be understood."

Jonathan and Helen, despite their age, knew that they were on the line to produce a creative response to the challenges that the entire class was now facing in their attempt to help Dan. If it were possible, their choice was to clarify the maturation process so that the therapeutic modalities might be successfully chosen on a basis of the patient's maturation and the awareness of whether inceptions (feelings), perceptions, or conceptions were more instrumental in the precipitation of the main symptoms.

With this in mind Helen responded. "I knew that we would eventually have to more carefully refine Dan's and Shang's therapeutic classification. I'm glad we're going to close these sessions with the class and clarify the method we'll have to use to decide which Alternative Therapies would be chosen for the Research Center. We've pretty well accepted the fact that before the Central Nervous System, the control of living functions was mainly under the control of the Endocrine glands. If we look carefully at the Endophilic Era, the major integrated control of the child's behavior is Endocrine dominated. This includes a dominant Thyroid gland, a moderately functional Adrenal Cortex of Mesophilic origin, and the less challenging Adrenal Medulla, which is of Ectodermal origin. It is the first example of how, during the Endophilic Era, the organs of Endodermal origin remained dominant unless they were failing in their functions, while the organs of Mesophilic and Ectophilic origin were slowly maturing and merely assisting or enhancing the functions of Endophilia. This is best seen with excessive stress during Selye's Adaptation syndrome.

It is these three organs that dominate in the formation of the energy, called Chi in Traditional Chinese Medicine, and during all the phases of Embryonal dominance, that precede the maturational changes that goals and behavior precipitate and bring into function. All of the Endodermal organs are functional in early life and that includes the lining of the pharynx including the Auditory tubes, the tympanic membrane, thyroid, tonsils, thymus, and parathyroids.

1) The respiratory tract, the larynx, trachea and lungs. It does not include the nose which is of Ectodermal origin.
2) Digestive tract, intrinsic glands, liver, gallbladder, bile ducts and pancreas.
3) The bladder.
4) The vagina.
5) Female urethra and the Bartholin glands.
6) The male urethra, prostate gland and Bulbo-urethral glands.

Every disease associated with Endodermal organ involvement is associated with a severe loss of energy. Consider the Hypothyroidism of a Cretin whose energy for growth is non existent.

In the evolvement process, as with all cells, it is the first to have a rhythmic polarization and depolarization that controls its function. Its rate is determined by the cell alone. In time, when it becomes part of a metacellular organ, it sacrifices its endogenous autonomy to the control of its organ system, which then takes over the pacemaking function of the entire metacellular organ. The organ is then in control of the rhythmic control of all of the cells of that particular organ.

In the same way that the organ takes over the rhythmic control of the cells, the human, by his goals and motivation, design his living systems and the Wellness or Illnesses he creates in his multi-systemic life. It is during this Endophilic era that the feelings regarding all activities, and the inner inceptions such as hunger, and the discomforts related to shortness of breath, become a powerful motivating force. Bowel and bladder signals take on the negative role that activate the urge to go to the bathroom, so that we learn that all of the signals of intrinsic origin that motivate specific behavior, are negative, and the specific behavioral responses to these signals are positive, so that at a very young age they are learning that they control the reactions to the negative signals in their life and it is in this control they experience the positive joys related to their actions. All of this occurs during the maternal-infant dyad, the most intimate and precious relationship that precedes all others.

Consider the magnificent consequences. As the infant feeds at the breast it feels the warmth of the mother, the smell of her breasts, and the taste of the milk, which is eventually followed by the sound of the mother's voice and the visual image of her face, all of which are associated with the assuagement of hunger, and its negation. It becomes the fixed image of the most positive feelings that the infant will ever have."

Jonathan stood in amazement as he listened to Helen. "Darling, I had no idea you had dug so deeply into the Embryology of Inceptual evolvement. I think I can guess where you're going with this analysis."

Helen smiled and continued her explanation. "During the first 2 years of life the Infantile Chaos experienced by the infant, if properly counteracted by an alert mom, always aware of her infant's emotional stress, will guarantee that a growing child will be wisened to his or her own negative feelings, and will successfully learn that even in the perceptual and conceptual world that will eventually require their analysis, they will be adding their own personal subjective coloring to most everything they experience and hear. It is not until they are nearly twelve years old that they begin to separate the subjective from the objective. Thus, just as the Embryonal phases of evolution shorten the true evolvement of mankind from 100 million or more years to only 9 months, the awakening of the subjective affect of man makes it quickly possible for each of us to classify, as good or bad, everything that is touched on by our perceptual apparatus for the rest of our lives. This also becomes part of our conceptual strengths as we mature into our adult years. This is the basis of human maturation as described by Eric Ericson. The power of the mother is at its apex at the moment of the infant's birth. It slowly dwindles with the passage of years. (An example of the autonomic growth, during the first two years of life, can be seen on pages 526 and 527 of the first Ortho-Para volume). For the rest of a persons life they are influenced by these emotional and feeling judgments.

Jonathan had listened attentively and clearly understood where she was going with her idea. He laughed as he responded. "I assume dear that you've designed the base on which the Mesophilic Era rests. Having designed the origin of both the negative and the positive feelings of every persons life, a very egocentric accomplishment, the base of the Allocentric Mesophilic Era is ready to be described.

All of the Mesodermal organs are immature at the time of birth and their functional maturity awaits their growth and an increase in their strength and endurance. They are as follows:

1) Smooth and Cardiac muscles
2) Connective tissues, cartilage and bone
3) Blood and bone marrow
4) Lymphoid organs (except the Thymus)
5) Suprarenal gland
6) Linings of the pleura, pericardium peritoneum, kidney, ureter, trigone of the Bladder, testes, epididymus, ductus deference, seminal vesicles, ovary, uterus, Uterine tubes, blood vessels, joint cavities, bursa, tendon sheaths, cavities of the Eyes and cochlea

These organs represent the beginning of the age of power. It begins at approximately the age of twelve with an upward refinement in strength, endurance, dexterity and skill. The nourishing environment of home has been modified by the challenges of school, sports, intellectual and physical competition, goals related to ultimate independence and work, the new awareness that they are reaching for adulthood and the careers and responsibilities that will fulfill them in their working years. The awareness of the self has been modified by the challenges that each system presents in relation to individual stature and the ability to survive in this world of challenge and opportunity.

What is the force that drives successful people forward? Beside a healthy supply of energy there has to be a powerful goal that is attached to a fantasy that if successfully reached will alleviate any negative feelings regarding all of the other systems of life. It is merely a fantasy and not a realistic goal that is ever reached. But it is not bad, even though unachievable. It keeps us aware that goals are essential to the living process in order to avoid the apathy and inertia that precedes severe depression.

Before we elaborate on the illnesses that dominate each era we must first introduce the Ectophilic era. The structures that begin to dominate the functional competence of this mature adult are as follows:

1) Skin, Sweat glands, Sebaceous glands, mammary glands, hair nails and lenses
2) Conjunctiva , Lining of the retina, lacrimal glands, external and inner ear, Nasal cavity, paranasal sinuses, mouth, teeth, taste buds, oral glands, Hypophysis, anus, tip of male urethra, adrenal medulla
3) Nervous tissue, including the neuroglia, chromatin tissue, and Autonomic Nervous system
4) Smooth muscle of the iris and sweat glands

The brain is divided into three distinct masses. The first mass is made up primarily of Rhombencephalon structures and the Mesencephalon, and is considered part of the Reptilian System. The older Mammalian Nervous system is made up of the addition of the Diencephalon which also contains the Limbic system of the brain. The forebrain or Telencephalon is considered the new brain, or Neocortex and it is that part of the brain that controls symbolic thinking. The neural control centers are as follows:

1) Spinal Cord
2) Myelencephalon or medulla
3) Metencephalon or Cerebellum and Pons
   Both the Myelencephalon and Metencephalon are considered part of the Primitive Rhombencephalon of the brain
4) Mesencephalon-containing specialized reflex centers
5) Diencephalon- Thalamoencephalon or interbrain
6) Telencephalon-Forebrain containing the Cerebral Hemispheres and Basal Nuclear Masses

These structures, inclusive of the nervous system, are slowly reaching maturity during the first 30 years of life. The ascending process of maturation is associated with a gradual increase of the Nissl substance in the cells that are slowly becoming functional, but the level of dominance must be high before this cellular response to Ectophilic constructive stress trips this functional, maturation response.

It becomes obvious that the Endophilic Era is and must be self centered and Egocentric because its power is the creation of those subjective centers designed to evaluate the worth, positive or negative, of the people, places and things that will become part of their lives.

The Mesophilic Era develops the power, strength, endurance, or skill to take on the variegated challenges the systems and particles of life will offer in every age of growth.

The Ectophilic Era will add the dimension of rational, creative thought as the power to carry on beyond that which the muscular phase of life will offer, so that the Endocrine System, the Autonomic Nervous System, the Musculo-skeletal structures of the body, and the Central Nervous System will participate in the responsibilities necessary to fulfill the potential of everyone who chooses to live their life to its fullest.

How do you translate the knowledge of all three treatises into a paradigm that fairly scrutinizes the potential therapies which desire to be a part of the Research center. That is the chore that we have undertaken to assist Dan in his designing of the Ortho-Para Research Center.

All the information for making a knowledgeable decision as to what professional healer is the most likely to successfully fulfill the needs of the patient are now available. The time has come to evaluate all of the Alternative therapies and to ascertain whether we can discover a paradigm which will guide us in determining what criteria we might use for helping us choose which therapies will become part of the Ortho-Para Research Center. The criteria would have to emphasize the Ortho-Para qualities of the Paradigm since the very goal of the Research Center is to introduce to the entire Medical Community to a new method of diagnosing and treating the Ill.

We can't over emphasize the role of the mother concerning her influence on the unfolding process during the first 6 years of life. She, and to a lesser degree, the father, play a major role in influencing the development of subjective biases during these early years of life. A partial explanation can be seen in chapter 6 (page 49). Every systemic tissue demand that requires a response from the infant and which can cause a serious discomfort that is relieved or heightened by the parents, will also fuse the action or perception to their subjective responses, so that any exposure to a similar circumstance in the future would precipitate a similar emotional reaction-positive or negative.

By 12, under these circumstances, an early adult with goals and ambitions is in the pre-phase of early Endomesophilia (early adulthood). All of the intuitive, irrational, passionate subjective, positive and negative feelings regarding God, Religion and Transcendentalism are now part of every answer to questions regarding the conceptual Ideas that underlie their approach to life. Before we present a complete and effective paradigm, we first have to present the data we're using so that a knowledgeable evaluation of the paradigm is possible. We'll attempt to list these first and then try to justify the ones we choose for the Research Center.

In order to first establish the Autonomic dominance of the patient they would first have to take the written Ortho-Para Test, a complete history, and a prolonged first visit with the evaluating physician. This will most likely determine the dominance of the patient, the dominance of the disease, and the most qualified practitioner capable of establishing a therapeutic relationship for treatment of the Illness.

To repeat: When a patient seeks help at the Research Center, he/she will be first seen by a physician in charge of an initial Ortho-Para evaluation. As part of the evaluation there will be:

1) A complete History- present and past
2) A general physical exam
3) The Ortho-ParaTest

4) A time consuming, relaxed visit with the doctor discussing the emotional content of the complaints

5) A tentative diagnosis

On a basis of the above, test confirmatory laboratory blood tests and x-rays will be performed. Then, on a basis of this evaluation the patient will be directed to an Alternative Therapist who represents one of the many Specialties who will represent The Research Center.

The information given to the treating doctor would encompass:

1) The results of the Ortho-Para test and a tentative Para/Ortho ratio. It would define the initial Embryologic Era that influences the patient represented.

2) It will suggest whether the Illness is either Ortho or Para dominant.

3) It will list the number of didactic and other systems in the patient's life and the education and jobs he/she's had and their apparent conceptual dominance.

Before we go on I'd like to clarify the meaning of what we'll now call the Para/Ortho ratio. We've changed the order of these dominances. We're born with the highest extreme of "Para" dominance we'll ever have, and it is only with the passage of time that the dominance of "Ortho" comes into existence. Although Ortho is present throughout life, as is Para, awareness of Ortho's powerful influence on Human behavior does not present itself until the age of twelve. This is the time of the weak Endomesophilic awakening of the Mesophilic Era. To clarify the ratio's presence, I will elaborate the relationship in all eras of the Embryonic classification.

Para/Ortho ratio of Endophilic Era
Endophilic Age-
9/1, 8.9/1.1, 8.8/1.2, 8.7/1.3, 8.6/1.4, 8.5/1.5, 8.4/1.6, 8.3/1.7, 8.2/1.8, 8.1/1.9

Mesoendophilic Age-
8/1, 7.9/1.1, 7.8/1.2, 7.7/1.3, 7.6/1.4, 7.5/1.5, 7.4/1.6, 7.3/1.7, 7.2/1.8, 7.1/1.9

Ectoendophilic Age-
7/1, 6.9/1.1, 6.8/1.2, 6.7/1.3, 6.6/1.4, 6.5/1.5, 6.4/1.6, 6.3/1.7, 6.2/1.8, 6.1/1.9

Para/Ortho ratio of Mesophilic Era
Endomesophilic Age-
6/1, 5.9/1.1, 5.8/1.2, 5.7/1.3, 5.6/1.4, 5.5 1.5, 5.4/1.6, 5.3/1.7, 5.2/1.8, 5.1/1.9

Wallace Salzman

Mesophilic Age-
5/1, 4.9/1.1, 4.8/1.2, 4.7/1.3, 4.6/1.4, 4.5/1.5, 4.4/1.6, 4.3/1.7, 4.2/1.8, 4.1/1.9

Ectomesophilic Age-
4/1, 3.9/1.1, 3.8/1.2, 3.7/1.3, 3.6/1.4, 3.5/1.5, 3.4/1.6, 3.3/1.7, 3.2/1.8, 3.1/1.9

Para/Ortho ratio of Ectophilic Era

Endoectophilic Age-
3/1, 2.9/1.1, 2.8/1.2, 2.7/1.3, 2.6/1.4, 2.5/1.5, 2.4/1.6, 2.3/1.7, 2.2/1.8, 2.1/1.9

Mesoectophilic Age-
2/1, 1.9/1.1, 1.8/1.2, 1.7/1.3, 1.6/1.4, 1.5/1.5, 1.4/1.6, 1.3/1.7, 1.2/1.8, 1.1/1.9

Ectophilic Age-
1/1, .9/1.1, .8/1.2, .7/1.3, .6/1.4, .5/1.5, .4/1.6, .3/1.7, .2/1.8, .1/1.9

With these ratios we have an adequate number of Ages in each Era to differentiate the periods from each other with representative conceptions which clarify the differences between a passionate, intuitive Para dominant period of life and a more logical, scientific Ortho period, when the essences of the child have been caste to the wind. When this happens, we look at the world through different eyes both physiologically, philosophically, and realistically, as described by Wilber in his book on the three eyes of Man. Physiologically, during the Endophilic Era, we are handicapped with the Para Veil and frequently see the world with the Eyes of the Spirit.

Discrete, epicritic, framed vision is not present and we see the world and the Universe in an egocentric way, believing that everything that happens will benefit us. It is at this time of life that truths and fantasies integrate into fantasies regarding the past, present and future and confabulation is felt to be a normal part of childhood.

It is during the Mesophilic Era that Philosophers rise to dominance. It is Wilken's contention that this is the result of seeing life through the Eye of the Mind. The analytic mind, so capable of justifying basic prejudice, comes to conclusions that are in accord with unverifiable beliefs. Wilber further states that the only solid truths are those that come, via science, using the Eye of the Flesh. It is untarnished with any prejudice and can be made believable to everyone because it is perceptually and measurably verifiable. This occurs most frequently during the Ectophilic Era, when Ortho dominance controls the thinking mechanisms.

We will attempt to relate the important concepts to the Embryonal dominances at the time they ontogenetically appeared. Since this is the first attempt to attach conceptual growth to an Para/Ortho ratio, there will be a lot of approximating.

Endophilic Concepts

1) Concept of the "I" being integrated with the entire Universe. It is before the differentiating process has begun- age-0-1   ratio-9/1-8.5/1.5
2) Concept of the world being responsive, via cries, to one's basic needs- age- 0-2   ratio- 8.5/1.5- 8.1/1.9
3) Concept of the first sign of language-
   age- 1-2   ratio- 8.5/1.5
4) Concept of anger as a means of getting attention-
   age- 1-3   ratio- 8.5/1.5
5) Concept of Dependency and Independency-
   age- 3-6   ratio- 8.1/1.9
6) Concept of appreciation for colors and music-
   age- 3-6   ratio- 8.1/1.9-7.1/1.9
7) Concept of hunger, satiation and negative refection.-
   age- 1   ratio-8.5/1.5
8) Concept of thirst and its quenching and negative refection-
   age- 1   ratio 8.5/1.5
9) Concept of gastrointestinal cramps /relief with defication.-
   age- 1-2   ratio-8.5/1.5
10) Concept of genitourinary pressure/relief with urination-
   age-1-2   ratio-8.5/1.5
11) Concept of togetherness and the feelings of separation from one's par-
   ents-   age-1   ratio-8.1/1.9
12) Concept of pleasure and pain- age-1-3   ratio-8.1/1.9
13) Concept of sexual awakening and its release with intimacy-
   age-12+   ratio- 7/1-6.1/3.9
14) Concept of spatial and depth perception   age-6-8   ratio- 7/1- 6.1/1.9
15) Concept of bigger and smaller   age-3-4   ratio- 8/1- 7.1/1.9
16) Concept of the many feelings that are associated with Ortho or Para dominance and how they influence our judgment-
   age- 6-12   ratio-7/1- 6.1/1.9
17) Concept of self worth vs inadequacy- age-3-6   ratio-8.1/1.9
18) Concept of being secure or fearful- age-1-3   ratio- 8.5/1.5
19) Concept of simple numbers-   age-3-6   ratio-8.1/1.9
20) Concept of size- age-3-6   ratio-8.1/1.9
21) Concept of global time-   age- 3-6   ratio-8.1/1.9
22) Concept of Public and Private-   age-6-12   ratio-7.1/1.9
23) Concept of meaning and the use of words, phrases and language-
   age-6-12   ratio-7.1/1.9

24) Concept of writing, drawing, and copying-
   ages-3-12+   ratio-8/1-6/1
25) Concept of Seductive power- age-3-6   ratio-8.1/1.9
26) Concept of Inductive thinking- age- 6-12   ratio-7.1/1.9
27) Concept of Egocentricity- age-3-6   ratio-8.1/1.9
28) Concept of Extensions; tools, instruments etc.- age-3-6   ratio-8.1/1.9
29) Concept of Spirits, Gods and eventually one God-
   age-6-12   ratio-7/1, 6.1/1.9
30) Concept of beautiful, ugly, trustworthy and untrustworthy-
   age 6-12   ratio-7/1
31) Concept of right and wrong, fearful and fearless-
   age 6-12   ratio-7/1
32) Concept of mortality, immortality, eternity-
   age-6-12   ratio-7/1

Mesophilic Concepts
   1) Concept of partistic time- age- 12-16   ratio-6/1-5.1/1.9
   2) Concept of physical powers; strength, speed, and dexterity-
      age-12-16   ratio-6/1- 5.1/1.9
   3) Concept of deductive thinking- age-12-16   ratio- 6/1 -5.1/1.9
   4) Concept of Allocentricity and the birth of the Ego-Allo conflict-
      age-16-26   ratio-5/1
   5) Concept of relationship between structure and function-
      age-16-26   ratio-5/1
   6) Newtonian concept of the Earth, the Universe, and Gravity-
      age-16-26   ratio-5/1
   7) Concept of Pacemaker-. age-16-26   ratio-5/1, 4.1/1.9
   8) Concept of integrated organ systems- age-12-16,   ratio- 5/1
   9) Concept of Evolution, Involution, maturation and regression-
      age-12-16   ratio-5/1
   10) Concept of Potential and Kinetic energy-   age-12-16   ratio-5/1
   11) Concept of Entropy and death; and non-entropy and birth-
      age 16-26   ratio-5/1, 4.1/1.9
   12) Concept of the Intuitive Whole leading to the Synthetic Whole-
      age 16-26   ratio5/1+
   13) Concept of the Irrational vs Rational (Passion vs. Reason)-
      age-16-26   ratio-5/1+
   14) Dualistic concept of Life and Thought-
      age-16-26   ratio- 5/1, 4.1/1.9
   15) Concepts of Monism, Dualism, and Relativism-
      age-16-26   ratio- 5/1

16) Concept the Three Eyes of Experiencing- the Spirit, the Mind and the Flesh- age-26-30   ratio- 4/1, 3.1/1.9
17) Concept of Inner Law, Social Law, and Universal law-
    age 26   ratio-4/1
18) Concept of Illness and Wellness- age- 26+   ratio-4/1
19) Concept of Communication and its many forms; energy emission, data collection, idea transference- age-26+   ratio- 4/1

Ectophilic Concepts
1) Concept of Special and General Relativity-   age- 30+   ratio-3/1, 2/1
2) Einsteinian Concepts- age-30+   ratio-2/1, 1/1
3) Concepts of Endophile, Mesophile and Ectophile values, feelings, and behavior based on the Embryonal concept of maturation of both Man and Society- age- 26+   ratio-2/1
4) Concept of Ortho and Para-   age-30+   ratio- 3/1
5) Concept of the reality of Living and Dying- age- 30   ratio-3/1
6) Concept of the Ortho and Para Veils-   age-30 +   ratio- 2/1
7) Concept of Cosmic Principles of Systems and Particles and a multi-centric Social theory-   age -30+   ratio-3/1
8) Concept of Ortho Societies (low context) and Para Societies (high context)-   age-30+   ratio-2/1
9) Concept of Contraction and Expansion-   age-30+   ratio-3/1
10) New Concepts of the C.N.S. with the slow Transformation of deep structure, its ascent, and its eventual Translation with maturation-
    age-26+   ratio-4/1
11) Concept of P.E.S. > P.E.C.> P.E.P.- age-26+   ratio- 4/1
12) Concept of the Universal evolvement from undifferentiated K.E., eventually leading to a Universe filled with both differentiated P.E. and K.E. age-30   ratio- 3/1

(note- It should be understood that the dynamics of these conceptual growths can be read in Ortho-Para #1- pages 351-360)

Chapter 31

Jonathan was feeling the pressure that Dan, and his requests for directions, had precipitated and was now ready to elaborate on the details to Helen. She sat quietly knowing that before long everyone would be advised regarding Jon's conclusions. He rose and began pacing. "Helen, today we have to begin and hopefully, successfully conclude our search for the paradigm that will help us decide what Alternative Medicines will be conducted in the Ortho-Para Research Center.

Much has been discovered in the lecture series given by our class of seventeen and they did a marvelous job in analyzing some of the different approaches to varied and sundry Illnesses.

On several occasions we succeeded in approaching an all inclusive notion. The first was our awakening to the fact that Para and Yin, and Ortho and Yang, reached similar conclusions, although they were based on totally different thought processes, 5000 years apart.

The second was the analysis of Dan and Shang, who approached the answer to our search by placing all diseases in three or four categories, some of which were extreme Ortho or Para, and their treatment was based on an autonomic deficit and therefore the use of a therapy that was the antitheses of the extreme.

The third was an awareness that the searching for the origin of Para was reaching back to that period when we were bathing in the primitive sea, the amniotic fluid from which we were born, and where pure Para and pure dependency on mom existed. The route out of the uterus, during birth, tripped an awareness of the subjective components of all perceptions that are programmed at the time of delivery and become the logical subjective route to once again experience the Para tranquility induced by sexual intimacy. Each perception becomes the focus of an Alternative route to healing, and the underlying process to all Alternative Medicine will depend on the preverbal perceptive subjectivity being awakened by the therapy.

This will become clearer as we describe the therapies other than Allopathic Medicine, so that Dan can complete his architectural drawings with some understanding. If you recall, we discussed this basic eureka some time ago, when I went to visit Becky and she introduced me to her physician Dr. Si Shu. To introduce the paradigm we should first list the percepts and concepts and the primitive subjective components of each that are awakened due to the therapeutic modality being used. We must remember that these are all programmed in the mind during the preverbal phase of infancy, so that a logical description of awakening can't be gained by psychologic techniques which use primarily the verbal areas of the brain. It will be easier to relate the awakening of perceptual subjectivity, which is mostly preverbal, to a primitive conceptual egocentric objectivity. This latter is somewhat dependent on being perceptual and verbal, and is related to powerful emotional impressions. The whole field of philosophy prior to the 20th century, rested on the bias of concepts that were programmed during the early period, when ritualized subjective ideas became a powerful part of thinking.

The physical part of Alternative Medicine is attached to feeling perceptions, and the Mental and Spiritual aspects of the same Medicine is attached to ritualized cultural, and System organized subjective thinking. They are mutually interdependent. It is important to understand that Infantile Anxiety may be the highest form of human anxiety and it must be dramatically appeased by the outside forces associated with maternal contact and the care taking chores with which she serves the babes reflex needs.

The therapeutic role of mom is imposed by the outside, in order to slowly return the babe to Para tranquility. She does so by imposing a goal oriented and less invasive Ortho technique to serve the initial but yet undifferentiated cause of the anxiety. Different attempts at serving the unhappy infant eventually teaches the mother the difference between hunger, thirst and a wet diaper, so that both mom and the babe have developed a means of communicating shortly after birth. The invasive maternalOrtho intensity decreases and guides it to a lesser Ortho and greater Para polarity as the survival therapy continues.

Thus Para therapy incorporates the severe Ortho of the birth process with a lesser Ortho, and in this way defines the Para as the contrast to the extreme Ortho it is treating. You do not incorporate a high Ortho with a high Para. It must be slowly and gently brought under control with a technique that is a lesser Ortho than that being treated and thus brought down via the perceptual magic route, to a Para position. In other words, the Ortho anxiety at first needs a lesser Ortho imposed by the outside world that has a greater amount of Para incorporated into the proautonomic extreme with its antithesis. You slowly lessen the pathologic dominance, initially with a lesser version of the same autonomic dominance.

The therapy best used would be that which can successfully encompass the existing dominance and slowly guide it in a therapeutic direction. The healers must present themselves with a friendly, if not loving, personality so that they're easily trusted. If accepted by the patient, they are most able to accomplish emotional miracles, so that all Alternative Therapies are somewhat dependent on the personality and dedication of the therapist and frequently not on the specific therapy being used.

We have noted that of all the most sensitive perceptual senses, and probably the most mature, is that of olfaction. A good part of the nerves to all the other perceptions are years away from reaching maturity because, like the eyes, they have to be exposed to the environmental stimulus to reach maturity. The nose is exposed to the amnionic fluid which allows it to reach maturity by the time of delivery. None of the other organs of perception are so tested before delivery. In this way the smell of the general vaginal secretions are programmed during the preverbal period and thus remain a nonverbal memory. This is an excellent reason for the Olfactory tract to be so sensitive and to be able to pick up the female pheromones and therefore contribute to the intrusive passions of an aggressive sexual male.

As we sit in the kitchen and smell the wonderful cooking, we generally note how powerfully we're overcome by the aromas that can either stimulate our appetites, or repel us. It is totally dependent on the culture and the programming of the infant during the early moments of life. You can believe that the smell of the lactating breast is the greatest smell of all in the infants world, and very likely for all of their lives. Perfumes and oils are used as powerful tools of attraction so that we daily witness the therapeutic positive effect of aromas as we live each day of our lives.

Thus we strangely enter the world with our first impressions that have some accuracy. It is how the small secretions of the vaginal vault effect our lives. Yet while the nose has so much power, our hearing and sight are still very immature and do not impress the newborn in the same way. They have to be exposed to sound and light before they develop diacritic perceptions."

Jonathan was patiently listened to by Helen, but it was obvious to Jon that she was straining to keep her silence. "You've been very patient, my dear, but you're straining to keep your silence, what's the matter?"

Helen laughed as she responded, "Because, my dear, there is more to be said before you go into your classification. But first I think it's time to break for lunch before we subject these wonderful people to any more lectures. When we return, there are several authors who should be heard before you continue. I'd like your permission to read you something I wrote five years ago."

Jon smiled back. "Obviously darling you have my permission. But first, we'll go to lunch."

They returned from lunch one hour later. After everyone was comfortably seated, Jonathan turned to Helen and asked, "Who was the author you wanted to mention before I started classifying?"

"I was reading 'The Psychologic Birth of The Human Infant- Symbiosis and Individuation' by Margaret Mahler & Group. She maintained that the infant takes shape in harmony and counterpoint with the mother, whether the mother is Healthy or Pathologic. No different than the configuration of a ligand and receptor, the infant is either designed rigidly and therefore firmly symbiotic, or with greater flexibility, so that the process of Individuation is easier accomplished successfully and with ease. At the time of birth the psyche is empty. It faces the turmoil and negativity of the birth phenomenon with a mighty Infantile Anxiety that is a kinetic fire taking place in and on a bundle of potential energy. Until the Para instincts awaken the infant to the negativity that recurs in inner life, and the outside world responds to these Para needs with positive assuagement, there is only negativity.

Thus birth, with the insult of the orgasmically imposed contractions of the uterus and vagina, the challenge of light, touch, dryness, coolness, external manipulation and then finally the absence of Oxygen which demands the first motivated effort that the first breath involves, we have good reason for our infant to think and feel, if he could think, that being born was a punishment. Then there is the loss of blood sugar that ushers in the feelings of anxiety, and a global combination of all inner feelings which have yet to be differentiated into what they individually ultimately become. Eventually they will learn what hunger is and what magical person will make the Ortho discomfort go away. All of this is in contrast to the Para tranquility that existed during our intra-uterine position.

But even as I say that, I realize that the intra-uterine recapitulation of phylogeny is being tripped by chemical stresses that precipitate the cellular differentiating processes that are not really different from the stresses that ultimately differentiate the newborn infant into the adult it will someday be.

In one sense the ontogenetic recapitulation of phylogeny is a capsulation of 3 billion years of environmental experimentation that contained the miracle of the sequential, catalytic and chemical environment that in recapitulation influenced the germ cell to process all of the primordia of life's phylogenetic past, which when encompassed, the newborn human child had been made.

In this sense, even before the infant is born to the stresses of life, its germ cell has already experienced the creative crises of all of the 3 billion years of Earth's gestational past, all within the 9 months of its magnificent own gestation period. Considering the physical insults of birth, it is to the benefit of the newborn to be so embryonally incomplete. In contrast to many other mammals, the infant human is atricial and therefore dependent, instead of precocial and ready to survive with evolution having completed the infant of that species. Whereas in the infant, the Endophilic organs that undergird the living processes are already working well, the organs of Mesodermal and Ectodermal origin are still incomplete and require an extrauterine environment and the perceptual systemic stimulation that are maturational signals for Meso and Ecto progressive evolvement.

The infant is therefore primarily an Endophile in Para dominance and in its first 2-3 weeks is dominantly autistic and still in the early phase of creating the symbiotic union of infant-mother. To the infant, the symbiotic union is a global union in which the mother is merely an extension of itself. Just as the Universe, in birth, was empty, so also the psyche of the infant.

The description of the infant and its mother takes on enormous importance as we try to explain the unfolding of the undifferentiated feelings, as they slowly differentiate from the anlage of infantile anxiety. One could call the initial undifferentiated feelings of the autistic one month old as being proprioceptive-enteroceptive and egocentrically cathected. As one approaches 2-3 months, with a firm establishment of the infant's awareness of the symbiosis between itself and the extension of itself (the mother), there is a shift to a sensori-perceptive cathexis, toward the periphery or outside world.

Although the sense of self requires the core of inner sensations, an introduction to the nature of the objective world (which is what the dyad ultimately represents) allows the inner needs to ultimately recognize their relationship to the outside world that serves them. The objective-subjective self is therefore forming in relation to the object world that serves it, but until the 7th month and into the 2nd year of life the crises of separation and individuation has not surfaced.

The author emphasizes that only when the body becomes the object of self love (narcissism}, due to the mother's long care, does the external object (the mother) become eligible for identification as a separate being.

During all of this time deglobalization and differentiation of the Central Nervous System is preparing the infant for that moment when coenesthetic perceptions begin to clear and simple images begin to develop outlines, clarity and texture. The normal autistic phase serves the postnatal consolidation of extrauterine Physiologic growth and promotes postnatal homeostasis. The normal symbiotic phase marks the all important phylogenetic capacity of the human being to invest the mother with the systemic powers in the soil of family from which all subsequent human relationships form. The separation-individuation phase is characterized by a steady increase in awareness of the self and the other self and the ushering in of the process of greater differentiation as the infant's inner and outer environments proceed in teaching the Ego-Allo lessons that must be learned with the passage of time. By the 4th month, the smile, the crucial sign of maternal-infant bonding, has taken place. During this period there are significant perceptual contacts that are programming the infant with essential physical contact with the outside world. In terms of human feelings, it is not what we see and hear that affects us as deeply and as inarticulately as that which touches us and that which we taste and smell, upon which can be grafted audible and visual images which can be as significant as the more intimate perceptions that titrate the negative enteroceptive and proprioceptive discomforts that holding, embrace, and the warmth of the other self can accomplish so quickly. The many feelings, that in their completeness we call love, are the dominant affects that not only occur during the infant-mother interaction but are eternally programmed into the affect perceptual receptors, so that we sense love by its touch, its feel, its smell and its taste and by the therapeutic responses generated in the body. In the growing infant it helps eliminate negative enteroceptive anxieties and when we've aged, the same inarticulate expressions of love are therapeutic to sick and pain filled bodies.

By the age of 4 months, when we're still in the symbiotic phase of development, we are already sensing those marvelous positive gifts that the mother, who is our outside world, gives us just by the way she holds, caresses and embraces us while we're being fed. Separation and individuation has yet to occur so that the shock of separation is ultimately far greater because of the awareness that the instruments of peace and fulfillment are really not a part of ourselves, but are part of our significant other, who must now agree to continue the same magic she has been performing since birth.

By the 6th or 7th months experimental separation has occurred frequently enough for the concept of individuation and separation to become a frightening and recurrently reaffirmed reality.

It is at this point that a stimulus for greater Ortho dominance arises and the un-channeled energy becomes a source for the anxieties and fears that become symptomatic at these ages. Naturally the autonomic polarity governs the great difference between the intrinsic responsiveness of the child to a rigid, cold or warm nurturing mother, or one who is distanced or close, or tactile, embracing, and cuddling, or any of the extremes when the infant is acting out its negative feelings. In this sense the intrinsic character of the infant is touching on the intrinsic character of the mother, and since there is such variability we can expect the end result of this daily encounter to produce all variations on the theme of normalcy and abnormalcy in the infant.

It is at this non verbal period of infant unfolding that the potential for verbal, tactile, olfactory, visual, gustatory, and the deep embrace used by the mother is programmed into the Limbic centers of the brain with both the negative intrinsic feelings of the child merged with the positive extrinsic feelings incorporating the child, so that once again inner inceptions are treated and cured by outer perceptions, controlled by mother". Helen stopped and looked at Jon. He was sitting with his eyes closed, obviously listening intensely and apparently liking what he heard. Helen continued, "I thought Jon, that you'd be pleased with her little evolvement story which agrees in every way with your new concepts of therapy. In my reading of Spitz, Mahler, Piaget and many of the other masters, I've seen no great disagreement."

Jon responded, "Since it's apparent that Dr. Mahler's assertions potentiate my birth description of perceptual preverbal powers, we can use a combination of both views regarding the origin of the intuitive powers of all perceptions. We have a logical, at least a legitimate reason to incorporate them into the Research Center.

For 5000 or more years, Love made children, and helped old folks get old and happy. Modern Medicine still denied its powers, and I was part of the denial system. The power of love and affection can no longer be denied as can now be seen by many of the Alternate Care healer representatives that have taken over the treatment of 40% of the chronically ill. The people have been convinced and they're bringing their purses to those men and women.

Now it's our job to research these fields properly and try to determine their true effectiveness. We've already described the Olfactory sense as the most mature at birth. It may be, but I would think the mouth and tongue are not far behind. Certainly the baby must think so.

Chapter 32

Jon and Helen were both happy with the basal Paradigm they had discussed and were now ready to elaborate on the Alternative Cares that were dependent on human perceptions. However they did not stand alone. They would build on infantile, Endophilic growth, up to the Ectoendophilic Age, at which time conceptual powers would have grown and be able to differentiate many human faiths, each organized and vested with the most powerful of all Alternative Healing mechanisms, the power of Religion and the church.

The first Perception we have chosen is likely the most mature at the time of birth.

OLFACTION (The Nose)

There is not a room in my home that didn't meet with the challenge from all of my faculties for final approval. So it was with my nose and its generous nerve supply that gave me pleasure in many ways. If these nerves weren't pleased they let me quickly know and whatever I had to do to alter that particular environment, I would go about making the necessary changes. This process was most likely to take place in the bathroom.

The kitchen was an olfactory Paradise when foods were being prepared and the bite of hunger, not necessarily present earlier, grew in intensity.

The bedroom smelled of cleanliness and fertility for it was there that the pheromones of woman initiated the intimate polarity of my entire body and set the stage, at times, for the pathway to physical love.

Every room, every closet had an olfactory memory that either drew me to it, or repelled me, usually because of dead organic matter. It was the first perception that matured, and the first that told me how the environment would treat me.

229

It was not the mortar or bricks. A relationship sometimes was dependent on an initial aroma. I was drawn by the women of my life, by their perceptual excellence. My eyes and nose never disagreed, so that my moments of intimacy were glorious moments of health and good feelings.

Why should the maker of such a life's essence be anything but the creator of health and the builder of a strong Immune system. The human nose is open to viral and bacterial invasion. It is constantly at war with the outside invaders that give us the minute doses of antigens that are necessary to build our immunity. It tests and taxes the immune system and if we were not protected by excessive antibiotics, our immune system might be more powerful and more effective than it is.

The nose, in its very aggressive ways, evaluates the unknown before any other close perception, and protects the others from any perceptual extremes. Now that we've elevated the egocentric powers of the nose, how can Aromatic therapy be therapeutic in treating sickness or disease?

To answer these questions we must understand the pathologic circumstances in which we might give Aroma therapy as advised by Aromatic therapists. Before we try to answer this question, when it comes to Alternate therapy I bow to Dr. Andrew Weil whose first degree was in Botany and whose bi-degree and present mode of practice gives him a more balanced faith in both Allopathic and Alternative therapies. He gave, what I consider, an excellent first recommendation regarding the choice of these two classification of doctors.

The Allopathic of Weil

1) All Trauma
2) Medical and Surgical Emergencies
3) Treat Bacterial Infections with Anti-biotics
4) Treat Fungus and Parasitic Infections
5) Treat Prevention with Immunization
6) Complex Diagnoses
7) Orthopedic Replacement Surgery
8) Cosmetic and Reconstructive Surgery
9) Diagnose and Replace Hormone Therapy

In terms of the Autonomic Nervous System they all are representatives of invasion from the outside world and would then be classified as Ortho diseases. In contrast, the second list under Alternative Therapists would be Para diseases caused by failures of the body to continue functioning normally. We will list these most frequent Endocrine and Autoimmune problems at another time.

Alternative Healers

?? Viral Diseases
1) Chronic Degenerative Diseases
2) Mental Illness
3) Allergic and Auto-immune diseases
4) Psycho-somatic (depressions and anxieties)
5) Cancer

The upper lists are Ortho diseases caused by the outside world. The unfortunate Evolvement of today's Allopathic Physicians is that they've lost or never learned the value of Touch and Intimacy in the treatment of patients. Para diseases require the same Love and affection as the newborn child, so that the creative processes can be restarted as they were in the early days following birth. It gives the degenerating organs a chance to re-fire their energies and thus perform their normal creative functions again.

One of the first functions of the maternal-infant dyad is the teaching of self Love to the infant. It is the responsibility of mom to love her baby so much that the baby finally believes, despite the severe discomforts of hunger, thirst and visceral pains which it doesn't understand during the early phase of life, that despite them, it is still worthwhile to love itself because everyone else does. This self love is the fire of kinetic energy that ushers in the ability of love to spread beyond itself. It not only grows with this energy, but it initiates creative maturation changes. It is the Love and closeness, the cuddling and embrace, that protopathically precipitates all of the Para creativity in the infant. It is the same polarity in the aging adult; the mechanism that perceptual endearment reproduces in the physiological processes when degeneration of Musculoskeletal diseases have not progressed too far. It is the power of functional regeneration to recognize the oncogenes that precipitated a cancer and is able to turn a serious problem around. It is a powerful suppressant of the depressing environment that brought on the cancer in the first place.

Remember, in every tissue there is a cellular create/destroy (Para/Ortho) ratio, and Endophilia with Para dominance can create more than we destroy. We seek in Alternative Touch Medicine to alter the Para/Ortho ratio, and change it to a positive whole integer.

There is no end to the conscious awareness of olfaction. When we're born, our first major crisis is caused by O_2 deficiency and we must suffer our first breath. The respiratory center in the basal brain causes a sudden stridor, respiratory gasp, and our life giving O_2 becomes part of life until the process of living comes to its final gasp. From gasp to gasp, we live our life of living and loving if we do it right. Our entire metabolism is based on an efficient O_2 supply which is equated to good health by most health experts. Our entire metabolism is based on an O2 supply that must be adequate if we're to survive with the sad Illnesses that strike when we're aged and most frequently have respiratory problems. O2, in its marriage to water, is the Para element. H2 is the intrusive Ortho element.

Since aging is a dehydration process, little else has to be said regarding the essential nature of O2 and its marital partner H2O.

I've had no practical or didactic experience in this Alternative therapy. I've been aware of the claims by those in the business of those precious oils which are used for Aromatic therapy and refer you to the book by Chris Hobbs 'Herbal Remedies for Dummies'. He is a 4th generation herbalist and wrote a chapter on Aromatics with a list of their functions.

List of use of Nasal Therapy

1) Aroma therapy and the Industry
2) Elements and meds for intubation
3) O2 therapy
4) Herbal therapy
5) Flower therapy
6) Environmental therapy
7) Dietary and food smells
8) Pulmonary and Gas therapy

Gustatory Therapy (The Mouth)

I've seen Echo pictures of my children in utero and on one occasion I could see the mouth open, and know that at that time it was busily bathed in that primitive Amnionic sea in which we float and swim and evolve so magically. I don't know if anyone has collected any volume of this magic sea and used it for therapy, but if it were found to produce magical and youthful changes, I would not be surprised.

We spoke of the needs for the perceptual organs to be exposed to the environment of their function before they can mature, and it was evident that the bathing of the mouth in amnionic fluid was evidence that the lips mouth and tongue were exposed to the maturing fluid, and all in utero.

We are talking about the organ that serves hunger, appetite, taste, delight, desire, thirst, hydration, dentition, chewing, salivation, and early digestion. In its lips we have a muco-cutaneous junction whose function in all of the perceptual organs is one of sensing passionate sensations, love and sensuality. All of the perceptual organs have this, and are the fertile place for hair to grow profusely. This also is so for the nose, mouth, eyes, ears, anus and genitalia.

If good health is joy and there is little that pleases us more than a wonderful meal and drink when our work is complete, our perceptual gifts are always around us and we have the choice of putting in our bodies that which is good for us or that which causes illness but makes us feel good by altering our normal chemistry and brain. In actuality it is these sensory perceptual organs that have major control over our life span, a fact that gives us some volitional control over our longevity.

In back of the throat are lymphocytic tissues, tonsils and adenoids, which have already been exposed to the contaminated vagina during the birth process, so that the function of growing Immunity takes place when we are coming into the world.

That powerful mouth and throat is feeding us, protecting us, pleasing us, and when in contact with our love object, she arouses our Para processes that ultimately assuage our love needs. It is the same process with man and woman when they make more babies.

All of these powers influence our babes even before they begin to serve the brain with chatter and communication. The Ectophilic steps of maturation can't be readied without language, the nidus of knowledge that has become the power of memory, culture and a communicative Society. We traveled through many millennia before the evolvement of the tools of language.

How does one successfully accept the burden of discussing the perceptual power of the mouth without talking about Nutritional Healing. It is not truly in the classification of Alternative medicine, but the average physician in Allopathic Medicine is too influenced by the market to be able to be an advocate for the best information. It is a huge subject, not completely taught even to the successful Nutritionists because their knowledge is changed, refuted and powerfully influenced by food servers who control the industry.

At one time, I began writing a book on Layered Man but he is layered in so many ways, I felt it too large and variegated a subject to complete. I feel, as far as the subject of clinical nutrition is concerned, I would recommend "Prescription for Nutritional Healing' by James Balch M.D. and Phyllis Balch C.N.C.

To determine what would be the best route to determine the ideal structure of the total body, would involve adequate knowledge of the dynamics of the layering process. It would answer many of our still unanswered questions regarding the dynamics of leading with the tools we must have to perfect our roles on Earth. Certainly that would give us adequate knowledge of the layering of Man, what he is, what he's made of, and what he needs to stay well.

In reading the works of men who attacked the making of life's structure, the real works of the men called Constructionists, and those who sought the same information by breaking down and analyzing structure called, Reductionists, I failed to find their work leading to a concept of healing via eating that is worthwhile writing about.

However to allow all readers to walk a similar path that I've been on, I decided to approach the subject of layering by comparing the Cosmic Creative evolvement process with that of Man and the Infant. A great part of that subject was already covered by you, Helen and it makes the discussion a little less complicated for both of us."

Jon stopped for a moment and saw Helen riveted to her seat in interest, thus he continued his enthusiastic speaking.

"Helen this will take me a long time. Some of it you heard before."

There were no complaints, so he began again.

"You'll sense the process and find it the best way to understand how complicated human motivation can be. (Similar material can be found in the latter Thesis in Otho-Para I).

Chapter 33

The atomic contents of life is represented in the first 25 elements of the Periodic Table- excluding the Noble elements. They are Hydrogen, Chlorine, Fluorine, Carbon, Bromine, Nitrogen, Iodine, Oxygen, Sodium, Magnesium, Aluminum, Silicon, Sulfur, Phosphorus, Potassium, Calcium, Iron, Chromium, Manganese, Copper, Zinc, Cobalt, and Nickel.

It is interesting that the composition of life is intermediate between the average composition of the Universe and the average composition of the Earth. 99%, both of the Universe and of Life is made up of 6 atoms. (Hydrogen, Helium, Carbon, Nitrogen, Oxygen, and Neon).

Is it possible that Life on Earth arose when the chemical composition of the Earth was much closer to the average composition of the Universe at that time and that some subsequent events have changed the gross chemical composition of the Earth?

Composition of the Universe.
Terrestrial Life, and the Earth's Crust (+/- 10%)

Atom	Universe	Terrestrial Life	Earth's Crust (+/-10%)
H_2	87	16	3
Helium	12	0	0
Carbon	.03	21	.1
Nitrogen	.008	3	.0001

Atom	Universe	Terrestrial Life	Earth's Crust (+/-10%)
Oxygen	.06	59	49
Neon	.02	0	0
Sodium	.0001	.01	.7
Magnesium	.0003	.04	8.0
Aluminum	.002	.001	2.0
Silicon	.003	1	14
Sulfur	.002	.02	.7
Phosphorus	.00003	.03	.07
Potassium	.000007	.1	.1
Argon	.0004	0	0
Calcium	.0001	.1	2.0
Iron	.0007	.005	18.0

In line with my direction of study, I began reviewing the Molecular Biology of the cell. At the time it was 33 years after my graduation from Medical School.

There were certain relevant surprises that amplified my knowledge of Evolution, as the author went into some detail regarding the origins and Biochemical functions of the Procaryocytes and Eucaryocytes. Based on the contention that the ocean of organic molecules that existed at the time of the beginning of Life, had little O2 and that the atmosphere also had little O2 until there was an adequate number of plants. The Procaryocytes had to have an O2 deficient metabolic system and gain their energy via the same processes that today's organisms precipitate fermentation in an anaerobic environment. This process, also known as glycolysis, is one in which glucose is degraded in the absence of O2, one that is still present in every living cell and drives the formation of ATP, the energy molecule which is used by all cells as a source of readily available chemical energy.

When O2 became available, other organisms arose which were aerobic in character and it postulated that the mitochondria were the small aerobic organisms with totally different DNA, which were ingested by the early Procaryocytes and that the origin of the first Eucaryocytic cell was via this process. Rather than be digested, they were symbiotically incorporated into the metabolic process of the new cell that could now have the advantage of having both anaerobic and aerobic processes for the creation of energy and therefore have the energy level associated with a high mobility and a more active metabolism.

Today, both anaerobic and aerobic bacteria exist, but by far the most dominant are the aerobic which have nuclei and mitochondria amongst many other intranuclear and cytoplasmic organelles that make Eucaryocytes far more complicated in their structure and more complex in their ability to survive and thrive.

The argument regarding viruses and where they fit in the evolving history of the cell has always bothered me. The question always remained what came first, the cell or the virus? It was a natural question. It was felt that viral replication and survival was dependent on a host cell, within which the virus replicated and in so doing destroyed the cell, and then via the fluids of the extracellular ocean the progeny did the same to many other cells. In this case the chicken came before the egg. But with the passage of time and the technical refinements of viral research, it became evident that the viruses varied in size, shape and in their genetic apparatus and that they very well have been part of the genetic experiment, as were the mitochondria.

Every virus has chromosomes with gene-like codes that are capable of directing the functions of one or more proteins and enzymes that assist both in their sustenance and replication so that the early experiments with life may very well have had these genetic particles floating in the primordeal sea where a myriad of miracles could take place, the viruses merely being one of the living particles reaching out to become the undergirding of the evolvement process that one day would become Man.

In telling our story regarding "Layered Man," we must go back to the primordial sea and learn a little bit more about the chemical porridge that contained the potential that was actualized when the first virus or cell awakened to its life giving responsibilities. We are talking about 3 to 3 billion years ago and about organisms that thrived in an anaerobic environment and awaited the eventual oxygenation of the atmosphere that set up the environment where the Eucaryocytes could thrive and redirect the metabolic cycle to one that was O2 dependent and via this route make possible the complex evolutionary evolvement that ultimately designed the primate and then Man as he is today.

As an aside, a virus is to a cell and eventually to Man, as a parasitic social welfare system is to a Nation. It designs a hierarchy that designs and encourages a group of people to take over part of the political apparatus of a Nation, and who then nurture these people with the ultimate end being National bankruptsy and chaos. Fortunately this was finally stopped, but we must be careful that Medicine and Social Security and a National Medical System doesn't end up doing the same thing.

What unanswerable question can we ask now? A tough one is, what was the elemental and terrestrial environment on day one when the lifeless Earth had finished its pre-life geologic phase and was ready to take on great challenges, using the aqueous chemical porridge of the sea to make the first life forms?

Since we know that the post Bang Earth was born 4 billion years ago and the oldest living organisms are approximately 3 billion years old that the diastrophic forces that were shifting the mantle layer and building the mountains and river valleys, had reached a level of quiescence that would allow a primordial cell to arise from the chemistry of the sea.

At least 92 elements had been made available to Earth by the stardust, produced by the stellar catastrophes that preceded the creation of our sun and its planetary system. As a third generation star, a good deal of the elemental generative process had already taken place within the galactic and its planetary systems of the stellar parents of the sun and what elemental building process that had yet to be completed, the Earth completed during its magmacious phase.

The environment in which life began was most likely in the estuaries of large ancient rivers where the silt bearing elements were sent down torrentially from their mountainous birthplace into a more quiescent place where the experiments of chemistry could take place. The environments could vary and it is quite possible that with the mantle of the earth being so thin during these days of early birth, the estuaries may have been heated up to boiling temperatures because we know that within hot springs that flow up from the depths of the earth that ancient bacteria called Archaebacteria dwell in heated abundance. Thus, as far as the medium where our elements experimented with pre-life bonding, they were hotly tested before the stable formula for permanent bonding to take place.

This test was no different than that which took place when the gestation of pre-matter was exposed to more than a million, million degrees and the mattergic contents of the Big Bang was forming the particles and energy that was birthed approximately 15 billion years ago.

The Particle Physicists talk about Threshhold Temperatures above which matter became energy and below which energy became the pre-atomic particles which one day became the inner structure of Hydrogen and all of the elements that followed. (This process is described in the Treatise at the end of the first volume of Ortho-Para I).

In the sea, and especially in a hot or boiling estuary, the gestating arm of the sea, the bonding powers of the elements were being fused and broken in endless experimentation, determining what mechanisms would ultimately light the spark essential for creating a creature that was not virtual, but substantial, and was capable of self sustenance and self replication.

As we know today, of the 92 elements that exist, only 84 are stable and not radio-active. Of these 84 elements, only 22 are essential to life and of those 22, 7 are needed in only trace amounts. The remaining 62 elements may have played an organizing role in the assembly of the complex megamolecules, or may in contrast have been the external test that the new bonds had to cope with as the precursor molecules of life were defining their functions. Sometime, probably 3 billion years ago a Threshhold Temperature was reached that allowed for both protein and lipid synthesis and eventually for the formation of the first membrane capable of enclosing and isolating its contents from the surrounding sea.

It is important to understand that the pre-biotic Earth was erupting with volcanos, lightening storms and torrential rains and an atmosphere without Ozone, so that the ultraviolet rays of the sun were participating in the biologic chaos that initially was not compatible with life. But time brought with it areas of concentrated elements that formed a soup that was layered with an H2S atmosphere. With the electrical energy of lightening and the bombardment of the Earth by Cosmic rays, the first beginnings of organic molecules using Carbon, Oxygen, and Nitrogen was brewing and the gestating sea slowly began to form Polymers made up of amino acids, nucleotides and with the high energy environment cooking up a storm, amino acids began to join to form Polypeptides, Nucleitides, and eventually Polynucleitides. The elements were creating molecules.The molecules were agglomerating to form Megamolecules and the chemistry of Life was being concentrated even before the first cell was formed. The Earth was a large chemistry lab and it was awaiting the formation of the first Lipoprotein membrane capable of folding and enclosing within itself this chemical soup that would one day become the spark of Life and would define the moment that the first cell became the progenitor of all of Life that was to follow.

Did this happen in only one estuary?

How many experiments were required before the spark ignited?

Were there many competent molecular newborns before one dominated, or was it also the beginning of the survival of the fittest and the beginning of conflict?

These are questions that can't be answered, but certainly the beginning of the experiment that created structure and function and was completed, so that 1 billion years after the Earth was completed Mother Earth's long pregnancy created the Procaryocyte, our first living cell.

The contents of a living cell is a compromise between the contents of the Universe and that of the Earth. A living cell is composed of a restricted number of elements, 6 of which are Carbon, Hydrogen, Oxygen, Nitrogen, Phosphorus, and Sulfur which surprisingly make up more than 99% of its weight.

Whereas the Universe is made up mostly of Hydrogen and Helium, the Earth is dominated by Oxygen and Silicon and the living cell's contents is a compromise between the contents of both.

Since Life began in the sea and was so dependent on sea water and its contents, it is apparent that the sea today is not reflective of the sea 3 billion years ago when the elemental contents were exposed to a very different temperature, atmospheric content and energy and a porridge rather than the solute that it is today. One can assume that the porridge of the primitive sea, minus the elemental and molecular content of all of life today, would give us the sea that now covers 2/3rds of our terrestrial surface today. Naturally, in the last 3 billion years of geologic evolvement, with the movement of the Continental plates and the shifting of Continents and seas, there are other reasons for the contents of the seas to have changed, but certainly with the huge amount of flora and fauna that cover the Earth, their bulk plays a large role in the changing contents of the seas. Since 70% of men and most creatures are made up of water and since our focus in this volume is to ultimately make comprehensible many of the Alternative therapies in Medicine, it is essential to understand the qualities of water that make it the Universal solvent of Life.(See the poem and chapters in the first Treatise in Ortho-Para volume I).

Although there may be many who recognize water as the most plentiful molecule on Earth, it is amazing to recall that it is made up of Hydrogen (an Ortho, and therefore a Male molecule and the most plentiful element in the Universe) and Oxygen, (a Para, and therefore a Female molecule and the most plentiful element on Earth). This is the end product of the marriage between the Universe and the Earth and makes up 70% of our bodies and 80% of our brains.

So much of the accumulated data on water can be found in "The Structure and Function of the Water Molecule" by Eisenberg.

It becomes apparent as we listed the qualities of water that its own enormous variability of function, even before it becomes a solvent within which other elements and molecules modify its actions, that water has electrical, magnetic and dipolar functions that can be influenced by some of the physiologic changes that take place during disease.

During Ortho dominance the edematous tumescence of the skin is minimal and it allows the sensitive, epicritic awareness of the environment that surrounds man. In contrast, during Para dominance the tumescence, although first only genital, soon encompasses the entire body and trips a protopathic awareness of both the environment and oneself. This change that can be both externally and internally controlled, markedly influences the behavior that can fluctuate between Reason and Passion, expansion and contraction, and even love or hate.

In identifying water as the first layer of man, we must be more sensitive to, in what way, the Laws of Systems and Particles touch on it.

The next layer of Man relates to the elements in water solution. It must be clearly understood that there is a dynamic interplay, as all elements move from their presence in the vascular sea to the extracellular fluid and finally to its cytoplasmic and intranuclear position.

It is the Hydrogen ion, known also as the Hydronium ion because it is carried on dipolar water in the form of H30+, that must initially dominate our thoughts. Because of the way it controls the activity of all the other elements and molecules, it becomes the dominant element of Man. In the process of Oxidation, it is the main by-product of the catabolic processes that turn structure into function, potential energy into kinetic energy. This occurs primarily when we are actively acting out our Ortho dominant behavior and acting out, striving physically to attain our goals.

In contrast, during the Reduction process, when we are rebuilding ourbodies, the Hydrogen is removed from its depot in the blood and extracellular fluid and once again becomes part of the structure of the body, only to await that moment when it once again becomes part of the energy. This occurs during Para dominance, mostly during the night as the slow alkalinization process reawakens the cells to their regenerative responsibilities.

It is the alkalinization of the egg by the entry of sperm, that begins the mitotic divisions of the egg. It is the alkalinization of the cell that takes place during the Para Reduction process that lifts the potential energy of the body cells and trips their regeneration in relation to the activities of the day before. That which was made acidotic due to catabolic oxidation becomes regenerative, with alkalosis, due to anabolism.

Thus the pH of the blood, extracellular and intracellular fluid, influence the build up or breakdown of the human, and is reflected in the concentration of the Hydronium ion in all of these fluid components.

If we look at most of the organic structures of the body, they are dominantly bonds of Carbon, Hydrogen, Oxygen, and Nitrogen but only Hydrogen controls and defines the Create/Destroy ratio of all the chemical structures of which we're made. The pH of balanced life is 7.45.

To detail the participation of Hydrogen in life processes, is the study of organic and physiologic chemistry and the molecular biology of the cell. The notations regarding its use in non-biologic processes, although not as extensive as in organic chemistry, is still irrelevant to our search for its role in life. Consider the fact that Hydrogen is part of the structure of glucose, its polysaccharides, triglycerides, fat, proteins and their polypeptides, and DNA and RNA and its polynucleitides. It is for this reason that biologic and synthetic organic chemistry is considered the study of hydrocarbons in a myriad of functional configurations.

Hydrogen is therefore both part of body structure, and body motion. It is also apparent that as we layer water and Hydrogen with the other elements of life, that water and Hydrogen will modify the structure and functions built into all of them.

What we've thus far described is a simple symposium on Hydrogen and its pH effect on human normalcy. It was presented to describe how complex the subject of Nutrition must be since we are involved with literally millions of chemical radicals and incomparably complex interactions. All of which, are dependent on the pH of the blood, which in turn is controlled by what we eat, what we do, what we drink, how much we move, and the countless changes in how we think, work, and are motivated.

For this reason, we are all distinct chemical, organic creatures. We come from different cultures, eat different foods, and react to them in relation to our present activity and our past genetic inheritance. We are more influenced by the power of the food and Pharmaceutical industry than by good science that truly understands what our body needs and how we can get these needs from the complex Flora and Fauna we feed on.

Undoubtedly we will move closer to a general truth about nutrition, but the conditions which increase Para cannot be revealed in analyzable or scientific terms.

Nutritional Healing has more to do with the mental state of the eater and how he happens to maintain a pH of 7.45 by the very way he mixes a life of Ortho with that of Parasympathetic Autonomic dominance, thus maintaining a normal Para/Ortho ratio. This life would indicate he was leading a fulfilling, well focused life style. In this way, the goals of Living and Loving, Reason and Passion, will perform their magic and allow for good health and longevity.

With this discussion fulfilled, our focus on the lips, tongue and mouth are finished. Before we attempt to generalize on the power of all Perceptual analyzers we will focus on the tactile sense of the skin.

Chapter 34
Tactility (The Skin)

It is difficult to imagine a primitive sea that has bathed an infant for 9 months and has magically accelerated 3 billion years of evolvement in 9 months. The Amnionic Sea, that intra-uterine fluid that had this magical affect on the nose and mouth could not have the same effect on the thick skin, so its Para effect was somewhat less powerful, but its power under certain circumstances caused both the Para internal and re-creative powers and the Ortho sustenance and protective powers to express themselves in the many ways we need for both the inner and outer demands of balanced living.

Throughout the writing of all 6 of my books, healthy, balanced living between Ortho and Para was found to be an essential ingredient of health. Traditional Chinese Medicine speaks of Yang and Yin in the same way and to an amazing degree Ortho and Para Medicine is a modern version of the same 5000 year old oriental brilliance.

The skin is a perceptual organ of Ectophilic origin and was designed to protect the rest of the body from destructive invasion. But it also was designed for constructive incorporation, and that proves to be the difference between its epicritic sensory protection against Ortho invasion, and the protopathic sensitivity and pleasure of Para incorporation. It therefore differs from the other peripheral organs thus far described, by taking into consideration the negative consequences of conflict and the means by which it can sometimes be handled with adequate information from the outside world.

Due to its bipolar sensitivity it offers a huge amount of data that must be constantly analyzed and is one of the minimal reasons why this analysis was written. It is the physician whose behavior in his practice responsibilities has altered the field of Medicine and whose attitudinal changes in relation to his patients has precipitated the overflow of Alternative Care Healers.

We are talking about the difference between Tech and Touch Medicine. Just as the nose with its hirsutism, mucous membranes, sinuses, turbinates and olfactory nerve tissues performs its olfactory responsibilities, and the mouth with its lips, teeth, tongue, tonsils, adenoids and salivary glands are accessories to deglutition, so also there are reachable structures, via the skin, namely the muscles, joints, tendons, bursa, and ligaments which are accessible via the many surfaces of the body and are the structures more amenable to Alternative Therapy than the skin itself. It is via the manipulation of these very approachable structures that an entire Alternative Industry has blossomed, and for good reasons. The Allopathic or Traditional medical delivery system failed to make itself cognizant of the therapeutic powers of closeness, passion and touch. It is this Touch Medicine, a Para delivery system, which we will focus on at this time.

Therapeutic Variables In Both Para And Ortho Alternative Therapies

Magic, Para, Non-Touch Therapy

Spiritual Healing
Color Therapy
Crystal ad Gem Stone Therapy
Cymatics
Radionics
Body Works
Rolfing
Hellerworks
Body Dynamic Therapy
Art Therapy
Dance Therapy
Music Therapy
Drama Therapy
Diagnostic Therapy
Kinethesiology
Scenery Therapy
Visualization Therapy
Hand Waving
Chanting
Charismatics
Divine Healing
Guided Imagery
Self Healing
Mesmeric Vital Energy
Psychic Energy

Botanic Garden Visualization
Psychoneuroimmunology
Humor Therapy
Psych Spiritual Education

Para Therapeutic Touch Therapy

Love Therapy
Hand Waving
Eastern Therapies
Acupuncture
Moxibustion
Auricular Therapy
Shiatsu and Acupressure
Chinese Herbalism
Ayurvedic Medicine
Polarity Medicine
Meditation
Autogenic Training
Psychneuroimmunology
Cognitive Therapy
Holistic Therapy
Aromatherapy
Homeopathy
Nutritional Therapy
Western Herbalism
Naturopathy
Bach Flower Therapy
Environmental Medicine
Dietary Medicine
Heat Therapy
Toxic And Hand Therapy
Hypo and Hyper ThermalTherapy
Relaxation and Visualization
Alexander Therapy
Hypnotherapy
Autogenic Training
Aston Patterning
JuiceTherapy
Micronutrition
Vitamin Therapy
Ortho Therapeutic Touch Invasion
Ophthalmology

Wallace Salzman

Light Therapy
Psychoneuroimmunology
Psychotherapy
Antibiotic Therapy
Resuscitation
Biofeedback Therapy
Osteopathy
Cranial Osteopathy
Chiropractic
Massage
Reflexology
Metamorphic Techniques
Body Manipulation
Bone setting
Reconstructive Surgery
Posture Therapy
Bioenergetics
Physical Therapy
Occupational Therapy
Cranio-Sacral Therapy
Organ Transplantation
Gene Therapy
Cloning
Pulmonary Therapy
Colon Therapy
Cardiac Therapy
Renal Therapy
G.I Therapy
G.U.Therapy
Vagal Stimulation
Myotherapy
Transcutaneus Nerve Stimulation
X-ray, MRI, Cat Scan
Dermatology
Ad Infinatum Therapy
Ortho Self Imposed Exercises
Yoga
T'ai Chi
Aikido
Aerobic Exercises
Quigong

"When I made this list, Helen, I was flabbergasted at the expansion that has occurred in Health variances in this country. This is probably not a complete list, yet there are 120 different approaches to Illness of the Body-Mind class of therapy that are available to the people in this country whose primary causes of death today are self induced and preventable.

We now have a list to classify diseases as to their Autonomic dominance. We also have a list of therapies that are classified as to Ortho-Para dominance and know that just as in Traditional Chinese Medicine your Yang Illnesses need Yin therapy, and in Autonomic diseases your Ortho Illnesses need Para therapies.

It is not the purpose of our study to delve into the diseases of Ortho and Para and to classify them, but we have the task of choosing for Dan, from the list, some representative Alternatives that will be carefully scrutinized and studied to determine their power to help heal an Illness or a type of Illness so that, in time, we can slowly evaluate over the years what may be a foolish scam and what has a clear validity to use under the proper circumstances. We know from the laws of Systems and Particles (to be found in the first Treatise) that contracting systems do not respond to rational or scientific study."

Helen was exhausted after Jon's long presentation but was curious as to how he would choose the Alternatives after such a long laundry list of professional candidates." What will you do Jon to decide from these candidates?"

Jon was drinking some water and was slow to answer. He then picked up the list before he spoke again. "Helen, it's still too early. I have much to say about the other Perceptual organs and also the Conceptual powers built on them. Obviously from the list, the structures that lie under the skin are most approachable and there will be quite a number of Mind-Body Alternatives that I'll choose, but we have yet to talk about the Ears, Anus, and genitalia which are relevant to our Perceptual analysis and certainly to the development of the influential Conceptual Powers that influence our health, our behavior and our lives."

Wallace Salzman

Chapter 35

Hearing (The Ear)

The ear and its External, Middle and Internal Meatus is a true fusion of all three Embryonal layers, with a muco-cutaneous junction that offers its profound sensitivities, a physical conducting system, and an electrical conducting pickup to multiple frequencies of sound waves that ultimately lay the foundation for speech, a container of knowledge and the creation of Conceptions that rest on the verbalizations of the Social System we live in. The tissue of the outer canal is enervated by the Para Vagal nerve, the all powerful conductor of Protopathic human hypersensitivities, and it is used by Acupuncturists in China to produce the miracle of auricular induced general anesthesia, sedation, and analgesia.

Out of sight and deep in the inner ear a complicated mechanism of translating air vibration frequencies into electrical frequencies takes place in the Cochlear. Until recently, if vascular insufficiently to the neural mechanisms precipitated a cochlear failure, nothing would be done, but today transplantation of a new Cochlear system has been successfully performed surgically and patients born deaf have a definite possibility of gaining hearing if the surgical transplant is performed before the hearing mechanism rises from the mid-brain geniculate bodies to the Neocortex.

Just as vision, the adult comprehension of sound depends on exposure to the frequencies of sound at a young age and if the transplant is performed too late there may be sound but it's never comprehensible.

The remarkable new Hearing Aids have helped people who thought they would never hear and with proper amplification they were no longer seriously handicapped.

The transfer of sound frequencies from the outside environment are very mechanical until the sound is transferred to the Cochlear and many problems

248

with the sequential transfer of sound waves from the tympanic membrane (ear drum), the middle ear three ossicles, the membrane into the inner ear, the Cochlear and the Semicircular canals, are all part of the exquisitely sensitive pick-up of a myriad of sound frequencies.

Since we're dealing with a complicated physical transfer of an environmental sound system, the major medical work is performed by Allopathic hearing Specialists, but if we review the list of Alternate Care Healers we find that the Alternate Care Para representatives have not neglected hearing as one route to Meditation, relaxation and sensuous pleasure. If only the words of affection used by two lovers was accepted by the listener, it would be considered a hearing Specialty that precedes the conjugal embrace which also precedes the building of the entire population.

The power of persuasion is based on language and hearing and one can easily say that all political goals of every system in this country are based and bounded by the mouth, its language and the ears. One could add to speech the power of music, the rhythm and tune, its pitch, its vibrato, and its intensity which influences the passions of certain people.

The following Alternative procedures can be considered attached to the ear.

Para
    Spiritual Healing
    Dance Therapy
    Music Therapy
    Diagnostic Therapy
    Chanting Therapy
    Divine Therapy
    Humor Therapy
    Etc

Ortho
    Auricular Therapy
    Psychoneuroimmunology
    Corneal Transplantation
    E.N T Specialty
    Etc

Wallace Salzman

Chapter 36
The Anus (Rectal Outlet)

We usually don't consider this Perceptual organ with thoughts of love, passion and affection, but in some groups or Societies anal play and intercourse is considered an acceptable part of love making.

To the average person the anus gives a kind of personal pleasure when it is functioning correctly with the passage of waste material. Since it is not under normal circumstances a subject discussed, it is naturally used for the comfortable or uncomfortable role of being used to evacuate.

The Genitalia (Love, Pleasure, and Infant Conception)

In chapter 6, the colorful description of the Epigeneticist's concept of the influence of the birth process and labor on the unfolding of the infant, I think would be wise to re-read at this time. No tissues in the baby are more powerfully controlled by Para Neural connections than the genitalia. As explained in the first Ortho-Para Novel, the process of intimacy is passionately overwhelming as the lovers recapitulate, in a short period of time, the convulsive journey down the birth canal. It is after a powerful Ortho personality has been subject to an orgasm, and its powerful post orgasmic tranquility, that the power of the shift from Ortho to Para can be experienced by all recipients of this powerful Autonomic shift.

In the formal anatomy and physiology presentation, we learned that the Para nerves are distributed by the Cranio-sacral group of nerves. The Cranial nerve is the tenth, and is known as the Vagus. The Sacral branches of this nerve system innervate the organs of the pelvis, and provide the protopathic ganglia close to each organ. It also controls the individual organs, which is in contrast to the Ortho nerve ganglia that are generously distributed farther away from the organs and initiate general body reactions rather than local organ responses. The process of rebuilding is specifically related to a local reaction rather than an Ortho generalized reaction. It is here that the differential of functions reveal themselves and the Vagal and Sacral powers can precipitate creative activities only in relation to the Ortho stress in the particular organ involved with the stress. A generalized reactive use of energy is needed for the unpredictable stress we deal with in the outside world, whereas the private, intimate rebuilding stimulus in the specific organ that has been Ortho stressed in the outside world, is a response being controlled by the local Para neural ganglia.

Certainly the reductive, creative process that takes place is best in an alkaline environment and requires an increase in blood that is best demonstrated by the erectile reaction that takes place in the penis of the male and the clitoris of the female. Following the vascular dilatation there is the edematous engorgement of the sexual organs and eventually the accessory organs of intimacy which include the breast, the skin that turns pink as the epicritic sensation of Ortho changes to the protopathic splendor of Para.

With the protopathic sensitization of the genitalia, the entire Mind-Set has a generalized Para augmentation and the polarity produces a tranquility whose orgasmic intensity is related to the intensity of the Ortho tension that preceded it, and is released.

The entire process is the last of all Perceptive organs to mature to a positive influence on the thoughts, behavior and internal aggressiveness, due primarily to the late development of Ortho dominance under hormone control. It is usually the male activity that maintains an Ortho dominance and is the reason why the male is more prone to respond to the orgasmic release of sexual intimacy. The female is more likely to be passive and to be seductively inviting the aggression of the male. The active behaviors may be different, but the by-product of the male and female contact is orgasmic for both if the male properly uses his powerful touch in foreplay to polarize the important pre-orgasmic Ortho that precedes a therapeutic orgasm.

We used the term positive and negative refection. The male, who is Ortho, tends to be hungry for a Para polarity. Perceptually satisfied hungers, whether it be for food, thirst, the look or voice of a child, a beer or sex, whenever the hunger is satisfied, it no longer is a driving force of energy. This reaction is considered a drive with its end in Para and is therefore an example of negative refection.

If the goal or motivation is to produce a creative product like a professional goal, a book, a big and challenging job, such as marriage, the requirement of energy is ever growing and is an example of positive refection because whatever you do, you will still have more energy in relation to the act and you will be Ortho dominant and remain in a powerful, well motivated mood. Para responses to Ortho driven behavior is considered a negative refective response. An Ortho response to powerful goal driven behavior is usually considered an example of positive refection.

Chapter 37

Conceptual Power and Evolvement

The Perceptual analysis so far given is a good method of understanding the subjective nature of our judgments regarding the people, places and things which we've experienced via these organs of Perception. It is certainly the way of understanding the nature and power of these Perceptions and in what way they create Ortho or Para feelings regarding that which we've seen, and the ultimate effect it can have on our health.

But there is also a Conceptual analysis that affects more than an individual. The power of a Concept in the hands of any person of power or great responsibility involves not the actions of just one person but of an entire army of people who are controlled, led and influenced by the Conceptual powers that remain the base of all action.

Understand that the relation of people to Concepts is similar to the Layering of physical Man through the cycle of elements, molecules, mega-molecules and eventually life forms. Consider systemic conceptual evolvement one that focuses on the evolvement of the individual, the marriage, parenthood, family, and all of the Social systems that rest on this group designed to solve Life's problems.

As a base to understanding, the Conceptual bases are embryonal in type. During the Endophilic Era, which for most people is the first 12 years, the concepts and the diseases will be Para in type.

During the Mesophilic Era, the maturation and type of diseases will be either Para-Ortho or Ortho-Para in type.

During the Ectophilic Era, if a person evolves to such a high level of maturation, he/she will be Ortho in Autonomic dominance and will probably be fighting Ortho destructive diseases.

Remember, Para involves creative, passionate, pre-rational thinking that's intuitive in type. It usually involves contracted private systems.

Ortho involves creative, logical and rational thinkers who are deductive and expansive in type and are usually reaching out to the social systems of life.

Remember the metabolism. Para is Anabolic, Alkalotic, Creative and controls the building of the body. Ortho is Catabolic, Acidotic, Destructive and turns structure into usable energy.

How can we classify these concepts so that we touch on the significant problems of the larger systems of Life.

Concepts
Religion
Nation
Culture

Religion, Culture and Nation

Since the dawn of language the concept of a higher Being has been part of the History of all cultures. By the time of the Greek Empire and the Age of the great Greek Philosophers, we see a gradual dilution of the significance of any one God with omniscience, and omnipotent power, and a change in the God concept which becomes systemic, with the various individual Gods living very Human life styles, with human frailties, in a mountainous Haven located on Mount Olympus. There is the romance of old Euhemeris, which relates to the Gods and the demigods of the ancient myths who are really men of preeminence, who were deified out of flattery or gratitude.

Historically, under the Egyptian Pharaoh, Anheptot created a concept of one sun God, but that concept did not fair too well with the priests at that time, and the one God concept did not return to influence the thinking of the masses until the story of Abraham and his flight from Mesopotamia at the order from God 1900 years before Christ. It was the beginning of Jewish Biblical History.

Hinduism, Taoism, Confucianism, and Buddhism became a significant part of a Philosophic religion during the Period of Philosophers about 600 B.C. The date strikes me, for some of the great Philosophers of Greece, namely the time of Sophocles, thrived around the same time. It was at this time that the brilliant concepts of Yin-Yang became part of the language of Confucius and Tao. Their Concepts of Philosophic free thinking caused a division of China into many Independent and mutually hostile States. This rather Ortho environment continued until 221 B.C., when the state of Ch'in conquered the last of its rivals and created the unified China, governed by a centrally appointed bureaucracy in place of the former hereditary landed nobility. In its contracted state, it became more Para, with minimal hostility from outside Nations for nearly 400 years.

Just as intra factional strife developed in China, following the Period of the Philosophers, there was strife in the city of Judea. It was a young Jewish Rabbi, Jesus Christ, who tried to influence some of the accepted religious behavior of that day, and the Rabbinic strife that ensued ultimately led to Political strife and the ultimate death of Christ who chose to walk this road of Religious sacrifice. This was not the first, nor the last, member of Judaism, or any other Religion, who has caused the internal division of every other Faith in the world. The birth of Christianity was merely another division of a growing faith and it did not become significantly different until the Pope of that time made Christianity the Roman National Religion in the year 200 A.D. It was just about 600 years later that Mohammed fought his way to establish the country of Arabia and the Religion of the Arab semites. To emphasize the propensities of Cain and Abel toward conflict, Isaac and Ishmael had sufficiently different concepts as to the use of land, and it also once again became the reason for conflict.

The one God, who ruled over the Cosmos, had created children as filled with the potential for hate as they were for love. Was the process supposed to create these antitheses? I believe that the history of the world, and my personal subjective encounters with power oriented people, would indicate that this belief is correct.

These ideas are not a tirade against the concept of One God. Throughout our conversation we demonstrated the origin of the concepts of a higher power and our ceaseless exposure to the awareness of the Perceptual and Conceptual power of mother associated with the subjective anticipations and their emotional rituals. Via ceaseless reinforcement, the existence of God changed from the intuitive to a passionate, intense belief that rules the lives of believers of their one faith. Sadly it is associated with dogmas that infer that each faith is in conflict, because not each one could possibly be worshipping the same God in so many different ways.

Apparently, the Perceptual notions upon which each Religious concept is built is not the same. The other interpretation of the conflict is related to the politics and the goals of all the priesthoods, which like political parties can gain and hold onto power, prestige and a large flock of parishioners, by offering not only a powerful belief in their God concept but an equally powerful bias against those who think otherwise.

The history of the men who evolved and helped mature each faith is the story of the birth of both the example of the unipolar force of Religious building, and the political nature of Religious political stragglers.

Remember Ortho is the energy of goal oriented energy and the birth therefore of competition and bias. It is the cause of systemic national good and also the frequent antipathy between Nations that do not share a common cause with their neighbor, so that the very same forces that underlie Religious antipathy can precipitate Cultural and National behavior that can lead to war and death. It apparently does not influence the behavior of the leaders who cling closely and tenaciously to their concepts of power and dominance. It is truly no different in our modern world with its so called intelligent and rational decision making. One can read the opinion and Will of the people of every faith and nation, and in speeches we hear the goal of Peace, and in the behavior and decision making, we see the makings of hostility, distrust, lies and the worldwide behavior that has led to systemic conflict since the beginning of civilization.

From this discussion, including the talks on Religion given by Dan and Becky (on page 178 Ortho-Para Novel III), we would appear to have a great distrust of the decisions of Man that were Para passionate rather than Ortho rational. If one came to the conclusion that we could not believe in God, there would be less than the many Religious denominations that make up most of the population of the world.

Since the goal of a good life is reaching for a balance between Living and Loving, living according to Social law and loving according to the instinctual pressures in one's private world, it is the integration of both that is the balance between Nature's Private world and the Public Social laws related to system power, Life and Longevity.

We must therefore understand the prejudicial, egotistic, primitive, passionate, subjective Inceptions and Perceptions upon which the Conceptions rest, so that we're able to eradicate the biases, prejudices and criticisms of a mental process that is mainly and always egocentric and not acceptable to the Para passionate beliefs of a neighbor who is judging them with a totally different passionate base, especially if an Ortho male is judging the notions of a Para female.

To attempt this most difficult mental process and successfully eliminate judgmentalism, despite the great difference in thinking processes, it would be very difficult and could possibly be done successfully only by a mature Ectomesophile or Endoectophile."

Jonathan was finished with his Perceptual and Conceptual analysis and quietly awaited a response from Helen. He had presented his arguments in lecture form and it took as much time to articulate as one of his Ortho-Para lectures at the University. Helen was sitting with her eyes closed but Jon knew that this was her posture for listening and analyzing the speaker's words, and Jon now anticipated Helen's response.

"Jon, if you remember that evening, several years ago, with Jeremy and Nellie. He had come to our home when Kenneth and Allie were there with the primary purpose of arguing about the laws of Fundamentalistic Christianity and the validity of the concept of Chosen-ness (Chapter 23 and 24- Ortho-Para Vol. III). For hours we discussed this analysis and when it was over, Jeremy accepted the decision of his son Kenneth to convert to Allie's Judaism, with an explanation that quickly answered the questions that you brought up in your discussion.

One need not logically explain the unexplainable in Para thinking, and need never be judged by anyone else. One merely expects others to not be the same, and if you are, you needn't bind yourself in a relationship bonded by a unipolar force against those whose intuitive Para thinking is different than yours. Different Perceptual and Conceptual bases will rarely be the same, even if they're brought up by the same mother and family. Since life evolved the process of differentiation, it is a process that never ends. We are not and never will be the same."

With this statement of certainty Helen had quickly summarized and ended the discussion of Inceptual, Perceptual and Conceptual powers in the creation of behavioral propensities.

"Jon, I think that the discussion of Traditional Chinese Medicine was adequate to the task of explaining Ayurvedic and Curanderismo Medicine, which is Nationally and Culturally ingrained in India and Mexico respectively. It would appear to me that the Shaman and the Witch Doctor had won the faith and confidence of their patients and therefore had a magical expertise, the most mystical power that every good healer was blessed with, and the power to perform the wonderful role of the physician.

I think the time has come for us to finish off the gathering of information for Dan. I think that Jason and Beth will finish off this rather long session.

All of the creative work that is encompassed in chapters 38 and 39 are with the kind permission of the Architectural firm of Syntech Inc, a Chicago and Glencoe, IL firm. It was with Steven Salzman's permission that his concept of the Life Group will be presented and hopefully it will become the incorporating notion, within which, the Ortho-Para Center's concepts will play a diagnostic and therapeutic function.

Chapter 38

Jonathan was pleased and amazed but the meeting had taken longer than he thought it would. There is still a great deal of work ahead and it's now time for Jason and Beth to clarify a few more therapies before we begin the difficult chore of choosing which of the therapies we'll be incorporating into the architectural plans for the Research Center.

This is an elusive subject that appears to defy reasonable classification and has therefore been a slippery subject to grasp onto firmly and to understand unequivocally. I suspected that it would be all of our problems, but you all must understand that a great number of these therapies were created during the Endophilic Era when the very processes of rational discrimination were just metaphysical theories without the intellectual know-how for changing inductive thinking, and its uncertainties, to deductive thinking and the scientific means of experimentation in order to determine a firm proof.

We must never forget that we were all initially pure Para Passion at birth and even as Ortho Reason surfaces, and in some dominates a great deal of our lives, we must still maintain an equilibrium between Passion and Reason in the way we live, if we are to stay well. The fact that Alternative Care does not fit into the scientific protocols of Modern Medicine may be the very reason for its success in the treatment of intrusive Ortho diseases. Jason and Beth, I hope you both can elaborate on this problem that we will all discuss fully when you're finished."

Beth wanted Jason to begin the presentation so that she might ultimately present the entire differential between the various approaches to the multiple polarities of Illness and Wellness Medicine. Jason stood up and walked to the podium. With a broad grin on his face he began, "Do you realize that this is the first time I have your complete and undivided attention without David or Laura butting in and disagreeing with me, even before I finish my introductory statement."

Laura immediately responded, "Just to make sure you don't feel that I love you less, let me make you more comfortable and let you know that in the last 25 years I've changed very little."

The class was in an uproar and Jason, still smiling, continued, "Thank you Laura. I knew that that was coming. Beth, you owe me two bucks. I knew Laura couldn't control herself." Despite the noise, Jason continued speaking. "My training in Sports Medicine placed me in a fine position to analyze all of the Body-Works therapies and I really don't believe they fit into the class of Alternative therapies since they are all merely modifications of the field of Physical Therapy. I have little concern that Dan will have difficulty designing space for any Body-Works therapies, especially since some of the hospitals are now making it part of their new departments.

I am therefore going to speak about Therapeutic Alternatives that have come into existence that are really on the border between mystery, magic and witch doctoring and maybe even charlatanism. I have several magazines that are being distributed at many weight reducing salons and health clubs that are not directly connected to the average hospital. They advertise seminars that are being conducted at conference centers throughout the country and to a great degree are offering ways to cope with the emotional problems of life that disrupt human balance and quite frequently are the causes of human illness.

Remember, with both Illness and Wellness, we are dealing with a state of mind. Like the swing of an emotional pendulum, our moods swing between Ortho and Para extremes and at both ends we can discover at times severe depression and chaotic anxiety. Needless to say they are both the undoing of organic integration and are the basis for immune incompetence.

Recognizing the power of thought and those who have the power to influence our belief systems, I want to mention some of the talents who are offering their services to the public. Immediately, I became aware that even in my own casual discussions with my clients, the number of different 'Mes" I can become are dependent on whom I'm talking to. To an Endophile, I talk the language of passion, mystery and magic, which is part of their belief systems. A

Mesophile may be in conflict between Passion and Reason, but Reason and Logic are very much a part of their understanding of life. I can approach them with more rational remedies that run parallel to their belief systems. The closer we get to Ectomesophilia and Ectophilia, the more essential it is that justifiable reason lay at the basis of our recommendations.

Since a high majority of the population of the world cling tenaciously to Endophilic beliefs, my discussions will revolve around that aspect of human growth and maturity and when I'm done you can all put me on the dissection table and cut away.

The first therapist I will introduce you to we'll call Yousif. He looks about 35 years old and he introduced himself as follows. He is a Vedic Astrologer; a Tarot Card Reader; a Psychic consultant; an Aura Photographer; an Aroma Therapist; and a Crystologist. Each one of his many talents touched on my reading in the last 10 years of my studies.

Vedic readings speak to the role of meditation and prayer that has, as its goal, conceptual and perceptual detachment from active life. It teaches that the Vedic study is the pathway to the eventual incorporation by God. It is in contrast to the Vedantic teachings of the Bhagavita Gita which insist that we all must live an active life so that we can evolve from the Tamasic stage of childhood, to the Rajasic phase of adulthood with a sense of responsibility to ourselves and others, and ultimately, with study, some are capable of reaching the Sattwic stage of Spirit and Intellect, and in that way fulfill all human responsibilities to both Man and God by overcoming the selfish resistances to maturation and adding something of worth to humankind.

The Vedic concepts in this country, no matter what your faith, will lead to withdrawal, depression and both physical and emotional dependency. The Vedantic concept is compatible with a normal, fulfilling life and is compatible with the struggle to balance the pendulous swings of the Ortho-Para pendulum.

Tarot card reading; palm reading, astrology, psychic consultants, aura photography and crystology have no place in treating the ill unless the patient deeply believes in the powers of the reader or psychic who, in this way, has the power of removing a negative hex imposed by a witch or charlatan whose goals were the symptoms that the patient truly believed were caused by his powerful magic.

We are speaking of the power of negative and positive thinking and the influence that both powerful benign and malignant charismatic people have on their friends or enemies. I couldn't help but believe that aromatherapy, in the hands of gentle, loving healers, might precipitate a healing Para dominance that might be beneficial to a patient who has a benign, Ortho precipitated, Illness.

I went to a metaphysical emporium which specialized in a myriad of gifts for the body, mind and spirit. Some of the classes that had good therapeutic potential were Belly dancing, T'ai Chi, Hatha Yoga, Meditation, and Massage, all of which are ritual motor therapies capable of precipitating relaxation and the balanced swing of the emotional pendulum.

Had it not been for the cost, I would have gone on a Caribbean cruise where I would have been taught Hypnotherapy and Neuro-subliminal Communication.

Not everyone can be hypnotized, which is a process that enhances subliminal communication. The more Para you are, the more open you are to control by another mind and the more likely to reveal inner thoughts that have been pushed into the subconscious, but are retrievable only with deep hypnotherapy. To my awareness, I have heard it used for the treatment of addictions, emotional and behavioral pathology that required, for its cure, the awakening of a cause that was submerged in one's unconscious. The Gestalt therapy of Fritz Perls was somewhat successful in accomplishing the same thing.

When I look at the advertisements of the myriad of Spiritual Healers, I think their powers can be covered with a general statement regarding all of the religious paths, their mystic offerings, and the many variations of sacred rituals that are part of every faith but more so in the ancient orthodoxies.

At the time I came to this conclusion, I realized the absolutely essential nature of a positive belief system and the power it can have for both the prevention and treatment of serious illnesses."

Jason's exposition was apparently finished and if not, Beth was bringing it to an end as she stood up and directed her comments at Jason. "Thank you Jason for giving me an opening for the final presentation of our class. Knowing that I would be the last presenter, I thought of some of the problems that Dan shared with me just after our last month's meeting, and I tried to relate our knowledge of the spectrum of Ortho-Para to the spectrum of individual and societal diseases that occur when either one of the polarities is extremely high or extremely low. In a sense, balance, that would indicate that the medical position of the Ortho-Para pendulum, determines the residence of autonomic health which, when in balance, is not a causative agent for any pathology. Intrinsically caused Pathology can only be precipitated by Ortho-Para extremes.

That does not however mean that the swing of the pendulum will not be moved to an Ortho or Para extreme by Illnesses caused by something other than autonomic extremes. There can be responsive changes to Somatic or Ortho Illnesses, rather than it being the functional causative agent of that being treated. But it is important to emphasize that whether autonomic extremes be the causative agent or a responsive factor to the Illness, it should be treated with equal enthusiasm.

The questions I gave myself were seemingly simple at the beginning and the more intensely I pursued the answers, the more complex the subject became.

In a sense, one's positive or negative attitude toward every specific system creates the Ortho or Para polarity that influences one's functional competence and therefore the behavior in the particular system involved. We are all born in Para dominance and into the systemic infant-maternal dyad that can help us unfold with an optimistic or pessimistic attitude toward our personal competence. If it is positive, we are encouraged to open ourselves more optimistically to new challenges in our family system and eventually into new systemic challenges in the outside world. If it is negative, we are moved in the opposite direction, become more self contained and approach outside challenges with more fear, less confidence and, at times, create failure even before the response to the challenge is complete. Thus positivity or negativity can be enhanced by the mother, the father and the older siblings in the family system. Negativity reinforces the normal birth Para; hinders the normal Ortho awakening and leads to childhood depression and all kinds of behavioral pathology.

Since functional behavior templates the structural changes that occur during Para nocturnal regeneration, the pathology of function can only create the pathology of structure, and functional competence slowly regresses along with structural competence unless the underlying negative intrinsic, autonomically controlled attitude is discovered and radically changed.

We start off life with a family, a systemic cheering squad that makes us feel the positive emotions upon which we can build a life of success, or the negative dynamics described above, and walk down a path of apparent inescapable failure. This was emphasized in Jonathan's second Treatise.

During every step in systemic growth our negative or positive attitude can be Modified, for better or worse; intensified or neutralized; reinforced or disabled, so that with each new systemic experience we find ourselves being modified. That system which allows us to grow and feel more confident, as we reach the age of independence, will be the one we move towards. We'll allow it to be the base of further growth as we take on the challenges that come with independence.

But we must remember that the first system, the maternal-infant dyad that had control during the differentiation of feelings (inceptions), has laid the base for either negative or positive feelings regarding one's self awareness, so that a man, husband and father, who has been parented by a mother who oversaw the creation of feelings of inadequacy in her infant, can still be controlled by this mother who maintains a level of feeling power in her child who has now become a dad.

In contrast, a well mothered infant and child can, as an adult, face many failures in life without losing the self confidence and power to continue to optimistically take on new challenges and eventually succeed.

I felt it necessary to emphasize Jonathan's belief that there is no more important time than our first system relationship that we all must enjoy or endure.

261

With this as the foundation to my discussion, I would like to challenge all of you, by defining a difference between Disease and Illness. You may all attack me before I'm finished, but let me go on. Disease is the structural invasion by the outside world and Illness is a functional abnormality due to our anomalous responses to inner systemic challenges.

I realize that structural invasion can cause functional abnormalities and systemic functional pathology will eventually lead to pathologic structural changes, but just as Ortho is an autonomic response to systemic challenges and invasion by the outside world, Para appears to be a functional failure of structure to contain and destroy the abnormalities of our inside world.

Illness is due to the failure of cellular responses to outside challenges, and Disease is due to the failure of human responses to outside invasion. You might think that the differential is slight and yet it ultimately influences the therapy needed to counteract the dire effects on life and the living process. To quickly and broadly differentiate the two, let me give an example. Pneumonia is an invasion of the structure of the lung by bacteria from the outside world. In contrast, if I failed an exam, it is due to poor preparation and dysfunction of my systemic responsibilities. The first is a poor cellular response and the second is a poor human response.

If you question the validity of this differential in relation to Illness and Disease, I give you as an example, the human dysfunctional behavior that precipitates most of the diseases that affect mankind today. Certainly the etiology of most Illnesses today are due to pathologic stress, emotional chaos, human imbalance, nutritional imbalance and in this industrial age, being bathed in the toxins we ingest and inhale every day.

The symptoms of Societal Illness are merely a reflection of dysfunctional man in his most precious systems of interaction. When you consider that 75% of Illness and Death today is secondary to life style toxins and stresses, we needn't have to look too far to discover the most obvious reasons why every optimistic social experiment should be looked at as a potential blessing.

We are not talking about windmill fighting. We are not building dreams on a bed of mercurial, substantiveless notions. Both the structures and functions that will be incorporated in our final product are the new truths of Wellness Medicine today and hopefully, to a much greater extent, tomorrow. They can have positive meaning for everyone, the young and the old, and for everyone who will be fortunate enough to be part of this exciting adventure in Health and Growth.

Having painted this dramatic sketch of our tomorrows, what evidences are there today that would indicate we might be ready to move in this direction?

1) There is a growing public consciousness regarding Health and Wellness.
2) Healthcare is being pushed out of the hospital setting and into Community health centers.

3) Hospitals are searching for new methods of serving their communities while remaining in the Medical business and in this way diversifying their sources of income.

4) Hospital Boards are agonizing over what their future role with their communities, their doctors, and their competitive hospitals should be.

5) Physicians are becoming more aware of Wellness Medicine.

6) Greater numbers of popular health magazines are attempting to teach people how and why to stay well.

7) There has been a significant increase in the lifespan. Is it to be filled with unhealthy or healthy people?

8) There is more time for leisure and recreation and a poor awareness as to how to healthfully use this time.

Although these are the positive signs that would indicate that the population is ready for some healthy changes in the concept of the role of the doctor, there are other signs- governmental, political, economic and professional that would indicate that there is much we will have to be moving away from.

1) Government and Insurance company reimbursement for Illness is being contracted.

2) There is an enormous increase in functional Illness- a symptom of individual and environmental stress.

3) Physicians are being harassed by every system of interaction with which they've worked so smoothly with before. It includes the patient, the family, the hospital, the state and federal government and the insurance industry. There are no systems that don't feel betrayed by our present technologic Medicine and the inordinate high cost of all of this care. The legal profession points its fingers at the physician as the cause.

4) Although alternative delivery systems are being implemented on an ever increasing basis, the consensus has yet to discover what the ultimate ideal system of tomorrow should and will be.

5) The structural fabric that might define the relationship between a Community and its Medical delivery system is still in its early creative phase.

As we examine the growing problems related to our Nation's health, we are faced with the answering of several difficult questions.

1) Can Technologic Medicine truly impact any further on the health statistics of this country without bankrupting us?

2) Is it possible that the investment in changing the environment, from one that is sickness oriented to one that is Wellness oriented, will impact on the National health care costs negatively?

3) One merely has to look at the morbidity and mortality statistics today to verify the fact that 75% of Illness today is self induced. If it is true, then

it is not what the doctor can do for the patient, but what the patient can do for him/her self, that is relevant to his life span.

4) The evidence as to why man is living longer and for some a healthier life, relates not only to the great successes of Medicine in the field of infectious diseases. It relates more to the personal and societal public health measures that, in the past, effectively eradicated the primary sources of epidemics.

These included better nutrition, purer water, better sewage disposal, better food hygiene, pasteurization, and with these we can add better jobs, pensions and the social programs for the elderly. The Immunization against many serious diseases has also played a major role in creating an environment where the causes for early death were markedly decreased.

We suggest that if societies can accomplish so much by having good health goals, goals that integrated the talents of an entire community, including the doctors that we should also be able to go beyond this and implement similar personal goals for everyone and make man less vulnerable to both attacks from without and within. If the community creates the potential for a longer and fuller life process, one that prepares its members to understand their work life potentials and can then assist them in the actualizing process, it seems plausible that this alone would make the community seductive to all who yearn for this route to an integrated life.

We've described our goals. It is time to put these primary goals aside and see if we are talking about something that makes good business sense in this materialistic world of ours. We can be idealistic dreamers but today, that which is implementable must also be profitable. This is the measure of worth in 2005 in the entire world, and I believe if we truly want to make practical suggestions, we must accept this reality. Idealistic dreams are valuable and possible but only if we choose to make them realities.

All complex systems form a hierarchy that can be likened to a Pyramid. We have chosen to call the complex Medical System a Pyramed. Vertically from above down and below up, all the levels of the hierarchy are of significant importance to each other. The realistic goals of one must influence the realistic goals of the others. This fact underlies the common bond that fuses them all, both vertically and horizontally if they are to be successful.

The vertical mortise of any Pyramed is the power to serve and command from above down and the need to be served and protected from below up. The hospitals and the physicians offer the people the power and skills of technologic and cognitive medicine within the tertiary, secondary and primary care settings. In return, with payments duly earned, the people give the medical profession the basal necessities of life, plus the isolatory cosmetics of power, a large home and more economic freedom. This, in turn, becomes the symbol of their worth to an appreciative and occasionally idolizing public.

The Medical School stands on the pinnacle of the Pyramed and offers wisdom and the potential answers to the more difficult of life's problems and the mysteries that lie still hidden within the body and the cell. This includes the answering of questions relating to the physical, spiritual and intellectual fears associated with Life and Death. In return the Medical Schools are given the isolation, the respect and the protection from the contentiousness of the public who are more prone to understand the practitioner of Medicine, but may find the researchers and their powers meaningless, mysterious and possibly dangerous because of their incomprehensibility.

To the practitioner, the researcher and teacher is a source of discovery of the new and profitable technologies that are sellable. To the Medical school, the rich hospitals and physicians are the source of large amounts of capital to make research and its associated high failure rate, possible. There is thus an interdependency between the apex and base of the Pyramed that allows each to measure life in their very own way. The apex is wisdom power. The Central core is economic and Industrial power and the base is composed of physical, emotional, seductive and spiritual power.

The horizontal integrative forces are different. They are more feeling and volatile and represent the bonds that bind marriage, family, neighbors and friends. When there is a common need, a common fear or a mutual enemy, the bonds are as powerful and as evanescent as love. When there is no apparent bond of mutuality, the most powerful force field of coherency is the cohesiveness of the laws of the system. These laws are judged and administered vertically by one horizontal level over another. Whether one examines the base, the core, or the apex, the absence of clearly defined reasons for bonded friendship is enough to precipitate the surfacing of a myriad of reasons for disliking and distrusting other members of the same level. In every system of relationship a pecking order of power and its potential anti-power, is always lurking in the background even when there is outward evidence of peace and friendship momentarily prevailing. The chaos of uncertainty that lies within each man is covered by a veneer of certainty and assurance. The moments of chaos within each marriage is covered by a veneer of friendship and hopefully an ever present affection. The chaos potential within each family is covered by a veneer of closeness and a union of purpose. When the chips are down and a source of danger arises from the outside, the outside world will only see this veneer. The egocentric conflicts that arise, like flash fires, within most systems, are quietly insulated from the observing outside world. If the dangers from the outside are great enough, they are insulated from each other.

This same stress exists within the Pyramed and remains part of the flammable characteristics of each horizontal level. It appears that every system, whether biologic or social, suffers from episodic chaotic disturbances when the common goals outside of the system are not organizing the participants of the system in relationship to the goals and each other. Either people join hands and unite in common purpose and in this way dispel both system and individual chaos and discontent, or they find themselves motivated by egocentric pacemakers, discover common discontents and create inner systemic chaos as they battle amongst each other.

If this truth is as biologic as breathing, and I believe it is, we must examine the question, "What's in it for me?" from a different perspective.

It has been accepted as a normal question to ask in the business world, which has its own special ethics and values, but in the professional world it has been looked at as the egocentric base for selfishness, cheating and aggressive, destructive social skirmishes and wars. The truth may seem disturbing, but the binding element of every functional system is the response to this question. Should you invest yourself in any system that gives you nothing of significance in return, whether it is marriage, family, work, church, or school? The answer is "No". There can be no Pyramed and no integrated society unless each individual can measure the system in terms of their personal gains and the price that must be paid.

As we look at this Pyramed we must discover the answers that relate to its goals, its powers and its sustained rather than evanescent viability. Note we are talking now about a Life Group Pyramed that lies within the greater Pyramid known as Nation and exists as the horizontal system that lies between the receivers of Medical care and the deliverers. What should be the product line that exists at each level of this Pyramed?

To The Medical School
1) A new and exciting Concept of Wellness Medicine.
2) An opportunity to restructure the delivery system by training doctors in the patient-teaching aspects of Medicine that are relevant to systemic relational dysfunction.
3) Involvement in the new and exciting Research and Development associated with Ortho-Para physiology and the system functions they serve.
4) An opportunity to legitimize the posture of Medicine and the Physicians in the field of Wellness.
5) The distinct possibility that this is the means of differentiating between the roles of the Primary and Secondary Physicians, the Generalists and the Specialists.

To The Hospitals
1) Tertiary care hospitals would play no roles in Wellness Medicine.
2) The secondary care hospitals would be the provider to the Community within which Wellness activities flourish. It is a legitimate means of cre-

ating a balanced service in which the Wellness functions subsidize the Illness functions that most hospitals now serve. To be certain, the people of the community must change their attitudes regarding the Medical industry whose goals and challenges must now be Health and Joy and a longer, fuller life.

To The Physicians

1) The word doctor in Greek means teacher. Although there are now many physicians who spend the time necessary to help people take care of themselves, it is not a service that is taught or emphasized in Medical schools or during one's training. For that matter the means of integrating the therapy for relational dysfunctions has never been presented in as complete a psycho-biologic package.

2) One of the inter-actional qualities of Medicine, that has been lost in the doctor-patient relationship, is the silent contract that guarantees to the patient a doctor's total dedication toward the resolution of his problems with compassion, energy and if necessary, without charge. The patriarchal mode of the delivery system that lifted this physician to the heights of a demi-God, also set him up for the expectations that sit at the base of most medico-legal suits today. The physician, as a teacher, can comfortably drop from the celestial seat, which has become so uncomfortable, and hand back to the patient his autonomy and the privilege of making himself well when such options are available. Wellness Medicine will prepare the physician to play this role and only when the tools of Traditional Medicine are essential to healing will he become the patient's assistant toward this end.

To The Corporate Employer

1) It makes available the means of keeping the working population healthy, with all of the necessary physical, emotional and Psychologic counselors and the accoutrements necessary to keep them all well, happy and fulfilled.

To The Unions

1) It gives the Unions the important goals for which to work besides the economic benefits that normally are the lobbying responsibilities of the worker's representatives.

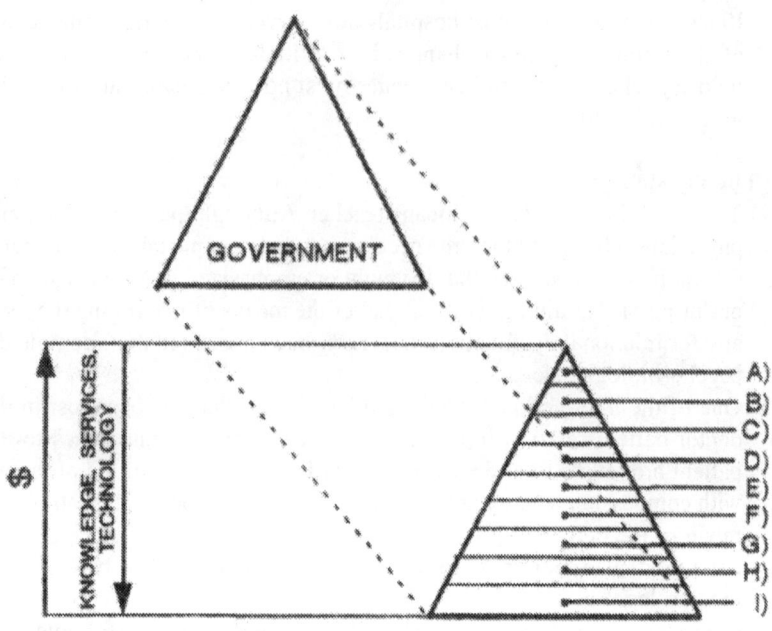

A) UNIVERSITY & PRIVATE RESEARCH

B) RESEARCH & TEACHING HOSPITALS

C) TERTIARY CARE HOSPITALS

D) SECONDARY HOSPITALS

E) PHYSICIANS' "PRIMARY CARE" CLINIC

F) L.I.F.E. COMMUNITY

G) HMO'S, PPO'S, INSURANCE

H) INDUSTRY

I) COMMUNITY (REQUIRED & DESIRED SERVICES & PRODUCTS)

L.I.F.E. PYRAMID

To The Patients
1) It gives them a Life Group that is designed to orchestrate the life style and health needs of its constituency. Since the autonomic polarities, propensities and life goals of each person are different, the expertise and responsibilities of the Life Group will be diverse and immense."

Jonathan sat there in amazement as he listened to Beth and after her last description could not control himself. "Beth please excuse me for interrupting you, but the amount of work that you've done calls for an explanation. Where did you get all of the data that you're presenting today?"

Beth smiled and responded rapidly to the question. "Do you remember meeting Nathan Ben Yosif?"

"Yes I do, but we're talking about several months ago."

"When he left you and after the sharing of your Ortho-Para theory, whenever he had the time and was not involved in his studies, he envisioned the day when you would be ready to influence the future changes in the Medical profession. He watched the training of Dan and only recently realized that Dan was given the architectural responsibilities of creating the structure that would house the Ortho-Para Research Center. I've known Nathan for years and when I shared with him the present goals of your project, he handed me a program which he thought you'd be interested in. I'll make a presentation to you regarding what Nathan called his ideas regarding a 'Life Institute Furthering Excellence', its acronym being 'L.I.F.E.', and with your permission Jonathan, I'd like to present a pictorial view of what Nathan had in mind."

Jonathan was startled at what his meeting with Nathan had precipitated.

"Beth, why didn't Nathan get in touch with me personally? Why did he do it so distantly? Did I offend him in any way? I honestly don't remember."

"No Jonathan. He didn't want to interfere with Dan's work. He just thought it would be wise to broaden your goals in a way that could eventually spread throughout the country and he was unaware as to whether you were thinking that big."

"Well Beth, I assume you've come prepared to let me know what program he is suggesting, so go ahead and we'll consider it as advice from a creative intellect."

Beth placed several documents into the projector that had been put in place even before the meeting started. "Nathan created a Logo for this Institute which he thought can eventually be incorporated under the corporate Logo of Ortho-Para. This, Jonathan, is your decision to make. Now if you all make yourselves comfortable I'll begin."

Wallace Salzman

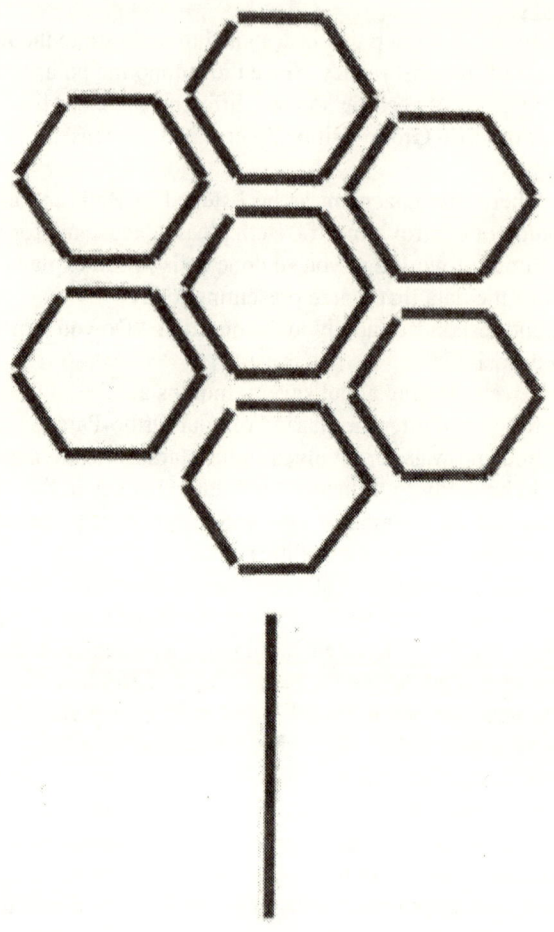

L.I.F.E.

Life Institute Furthering Excellence

The Life Institute For The Furthering Of Excellence

L.I.F.E.

There are two parallel and concurrent phenomena in our Health Care System that are evidence of changes in our Modern world. The first is that some of us are Living longer and that our chances for living full and healthy lives are far better today than it was 50 years ago.

The second is that some of us are Dying longer and that our chances for dying via a prolonged and debilitating illness are much higher than they were 50 years ago.

The very lessons that those who are Living Longer teach us become the base for the balanced and integrated programs that underlie the principles of the Life Institute for Furthering Excellence. In understanding and structuring an environment within each Community that offers the children, adults and elders a place to act out those essential, health oriented activities that promote good feelings, high energy and positive notions, we are promoting the genesis of both individual and communal healing and good health. The very positive gains that are experienced individually will make the joining of hands, that is essential to sustaining a community project, an enthusiastic and willing commitment by all who have been nurtured by its presence.

We must remember that L.I.F.E. blossoms within an environment that nurtures trust, encourages play and fun, and offers tasty platters of learning that can precipitate the growth and unfolding essential to the making of independent, productive and giving individuals. It is their physical, intellectual and creative activities that will make up the healthy communities of tomorrow.

Nathan feels that the entire concept of the Life Community rests firmly on the activation of the Ortho-Para theory in the study of a patient community. The community would have the physical environment, the programs, the products and the skilled professionals that are necessary to serve, to teach and to guide the people of the participating communities to move toward those mutually positive ends we've described. The Wellness Medicine of tomorrow, that L.I.F.E. represents, will add the services of the teacher physician whose responsibilities will be to teach his students how to take care of themselves in such way that the students will eventually be able to take over the teacher's role with his/her family and as this knowledge grows, the need for the primary teacher, the physician, will get less and less.

Today's technologic and medical miracles have played an important role in expanding the average life span of our population, but along with this gift of life we now see an increasing number of ethical and moral dilemmas that relate to the questions regarding dying and the cessation of therapy. Hopefully, via the serious dialogue taking place at many levels today, the decisions regarding 'when invasive technologic therapy must stop', will be defined more clearly and the tragedy of unnecessary and painful dying will be remedied. Within our L.I.F.E. community, both Illness and Wellness Medicine must remain married to each other, each having their own special facilities and job description. In this way the community will remain an integrated system, in balance both functionally and structurally and by all means both economically and ethically. It will be the jewel of every community that has it, and it must be sufficiently flexible to dynamically alter its directions when new conceptual changes make it desirable.

The Health Care Issues

Today's health care environment is filled with many difficult and agonizing questions. There are many driving forces which provide the impetus for change. A victorious health delivery system will be able to successfully accommodate each of these issues.

Cost

1) There is a dramatic increase in the cost of medical care in this country that has touched hard on the Federal budget because the government is Medicine's biggest customer.

2) This increase in cost is associated with an increase in those who are uninsured. They have become the 40 million people who get their medical services delivered in the Emergency Rooms of this country, the most expensive delivery system. The burden of payment is too frequently laid on the shoulders of the Federal government or the Hospital itself.

3) Technocratic growth in our nation has produced technologies that are amazing in what they do and terribly expensive to deliver.

4) The Pharmaceutical Industry has become a hungry giant that is eating up a good share of the medical dollars because of the very high prices being charged.

5) Partially, because of the improvement in Technologic and Therapeutic knowledge, we are living longer but at a huge technologic cost attached to the dying process of our elders.

6) Research in fertilization technology and the associated Genetic research has become a new and very expensive Specialty. We can anticipate that the products of the Genome project will increase this many-fold.

7) Tort law adds a huge multiplier to the cost of the delivery of every service that physicians and hospitals offer to their patients. The very fear of malpractice has designed a defensive medical system that over-tests, and at times over-treats, to assure that legal vulnerability is minimized. This is the Defensive Medicine that is now taught in Medical Schools.

Rise in Health Care Awareness

1) There is an increase in the amount of time for leisure and recreation for a growing population who have little awareness of how to healthfully fill this time.
2) There is a marked increase in the concerns for balanced health, but a scarcity of integrated professional programs to teach people the cause of their problems and what to do to stay well.
3) There is an enormous increase in Mental Illness encompassing the full spectrum of human feelings, from the severe anxieties to reactive depressions. They both cause immobility and are the seed of destruction for both the work and family environment. Look to the inner city for confirmation.
4) There is a large population of Elders who do not know if a longer life is for Living longer or Dying longer.
5) There is a rise in consciousness regarding the worth of a healthy body, a healthy mind and balanced serenity. These incentives are now joined by an awareness that:
   a. Medical care is moving out of the hospital setting.
   b. Physicians must be more involved with Preventative Medicine.
   c. Staying well will prove cheaper than getting well.

Social Disintegration

To the above, we must add the huge issues regarding mental, family and social disintegration. We have been so involved with casual, superficial networking that serve our egoistic needs that the benefits associated with Nuclear family have withered away and the mortise that binds the past to the future, our Elders to our Children, has withered with it.

With one parent families, especially in the slums of the inner city, the absence of family bonding has been replaced by gang bonding. With an extension of the social conflict that appears present even at the highest levels of our society, verbal and conceptual wars have been turned into the wars with guns and knives.

Ethical and Patient Issues

1) What is appropriate Health Care delivery today?
2) How much should I pay for the services?
3) What should I expect to get for services rendered?

The Conflict between Natural Law and Social Law

There is an interesting quiet war going on that has touched on the benevolent relationship between the doctor and patient. When technologic advances placed the physician in a position where he could control the dying process and via extraordinary resuscitative techniques revive those who were previously

273

considered dead, the legal profession saw the potential surfacing of a new social phenomenon where their skills, as the protectors of Social law, came into existence. Were physicians using their new knowledge properly?

Death, which was once in the province of God alone, was now in the hands of the physician.

Thus Malpractice law, which was once a very small branch of law, grew into its present state. It designed the rise of malpractice attorneys, malpractice premiums, malpractice suits and as a result the higher cost for each patient who was burdened with this new professional service.

The rise of new technologies in Radiology and the introduction of Endoscopic surgery in all of the specialties precipitated a movement away from general Medicine and toward an increase in the number of specialties and sub-specialties. The cost of Medical services rose accordingly.

The medical field slowly lost its benevolent and elevated position amongst the general population. Conflicts between the legal and medical profession created one reason for the rise of distrust. Conflict between the generalist and the specialist created another source of strife. The wars between the hospitals for service areas and the negative marketing being used to undermine the reputations of their competitors, revealed the destructive weaponry that has been increasing the distrust that the average citizen now has for the Medical profession. Add to that the burden of cost, imposed by both hospital and doctor, as they cope with the burden of paper work imposed by Federal and State mandates and you have a clear picture of the many factors involved in the declining adoration of the physician and his profession.

Introduction to the L.I.F.E. Concept

The L.I.F.E. Concept touches on every issue that has thus far been presented. It represents a long commitment to designing a Life oriented Community that must become the environment within which the problems of the Medical Delivery System can be successfully resolved.

It will introduce a new system focused on Health and Wellness which encompasses some of the fitness programs prescribed today and then goes far beyond. It presents a new approach to functional Illness that is behaviorally oriented. It will be less dependent on the eternal use of mood altering drugs.

It will provide both high tech scientific care and high touch compassionate care, both being essential to the balanced offerings that every good medical system must encompass. It will extract that which is best in the Illness delivery system and marry it to the new Wellness delivery system that must slowly rise to prominence if the cost of Medicine is to drop and the emphasis on healthy living is to supplant the present focus on Illness and Dying.

It will provide the mechanisms for Research and Development in both the fields of Illness and Wellness Medicine, hopefully shifting the emphasis to Wellness Medicine. We define the relationship between the doctor and patient differently in these two areas of the Medical profession. In Illness Medicine the physician asks the question "What can I do to help you?" The dependent relationship becomes at times eternal. In Wellness Medicine the physician asks the question "What can I teach you so that you understand how to take care of yourself?" It slowly changes the student patient into a wiser advocate of health knowledge. One who can become a teacher within the family unit and decrease, in this way, the teaching load of the physician.

Philosophically, pragmatically, politically, morally and economically, it is imperative that we create a unified system capable of implementing the goals of health we've so far described.

The L.I.F.E. Concept brings into existence the following:

1) The L.I.F.E. Group- as both the guide and participant.
2) The L.I.F.E. Community- the main goal we must achieve.
3) The L.I.F.E. Program- the design, the process, the philosophy and the business plan that can make it all happen.

The L.I.F.E. Community

The Medical Facilities of the future will offer a full range of L.I.F.E. services to the Community of clients and patients. The L.I.F.E. Community will be the staging ground for the implementing of both Patient care and offerings of the knowledge and facilities essential to balance health care with the maintenance of both structural and functional Wellness. Its goal is to promote a rebirth of the positive image that Medicine lost nearly 30 years ago. The consumer, who is part of the system, will begin to understand that the representative professionals of the Health Care Field are concerned about them and offer services that confirm and reconfirm that posture with every contact.

The Eight Elements that are
The Foundation of the L.I.F.E Community

Daycare	Health Care	Diagnostic Care
Infant	Family Practice	All High Technology
Child	Internal Medicine	Non InvasiveProcedure
Adult	Obstetrics and Gyn	
Elders	Ophthalmology	
	Dentistry	
	Primary Care	

Eldercare	**Urgent Care**	**Educational Awareness**
Apartments	Emergency	Lectures
Congregate	Ambulatory	Seminars
Life Care		Health Products
Reactivation		
Skilled Nursing		

Discover Care	**Hospital**
Swimming	The Community Hospital
Physical Therapy	
Aerobics	
Programs to enhance Health and Wellness	

Discover Care

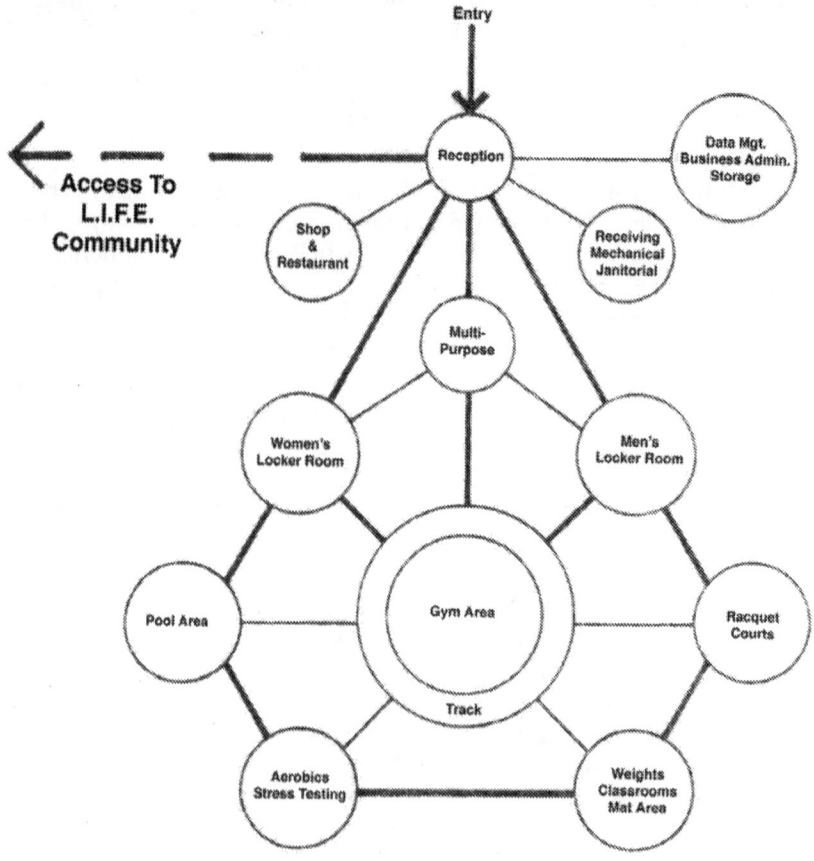

Elder Care

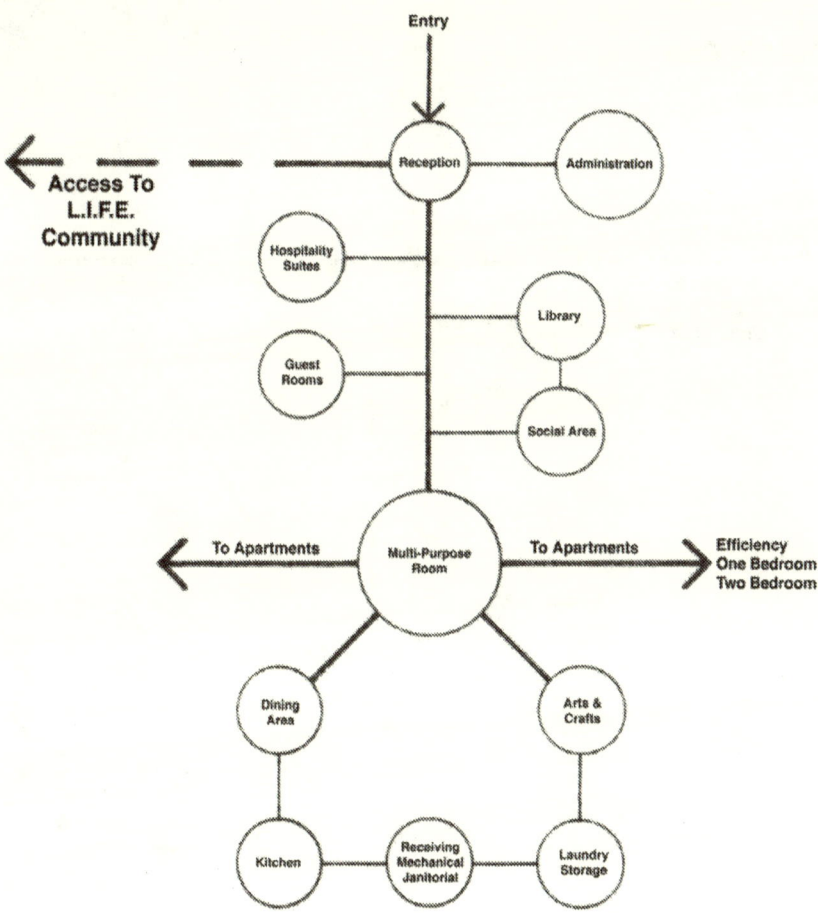

Wallace Salzman

Health Care

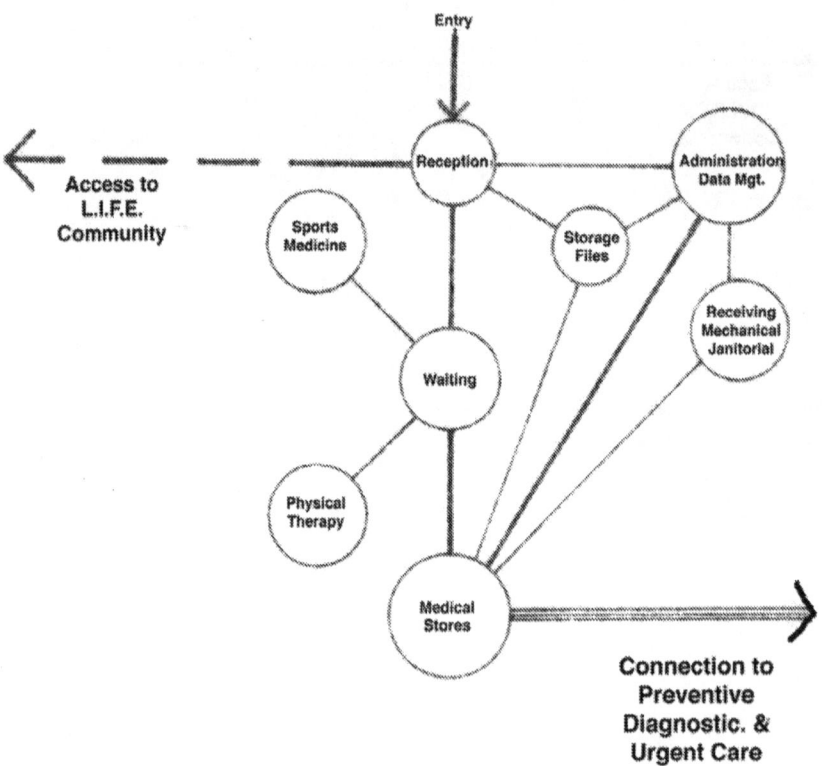

278

Diagnostic Care

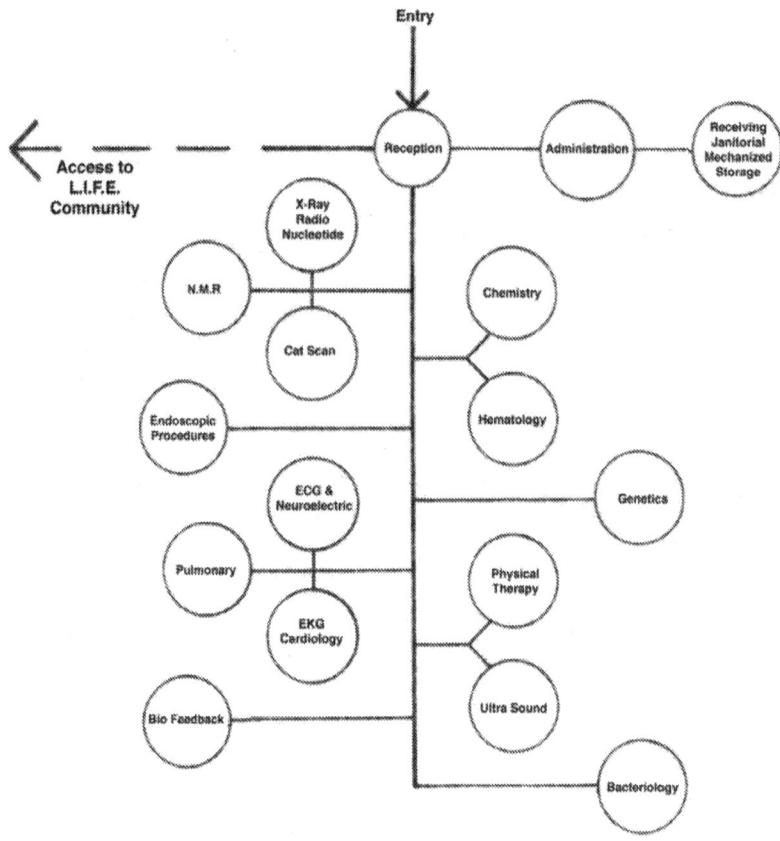

Day Care

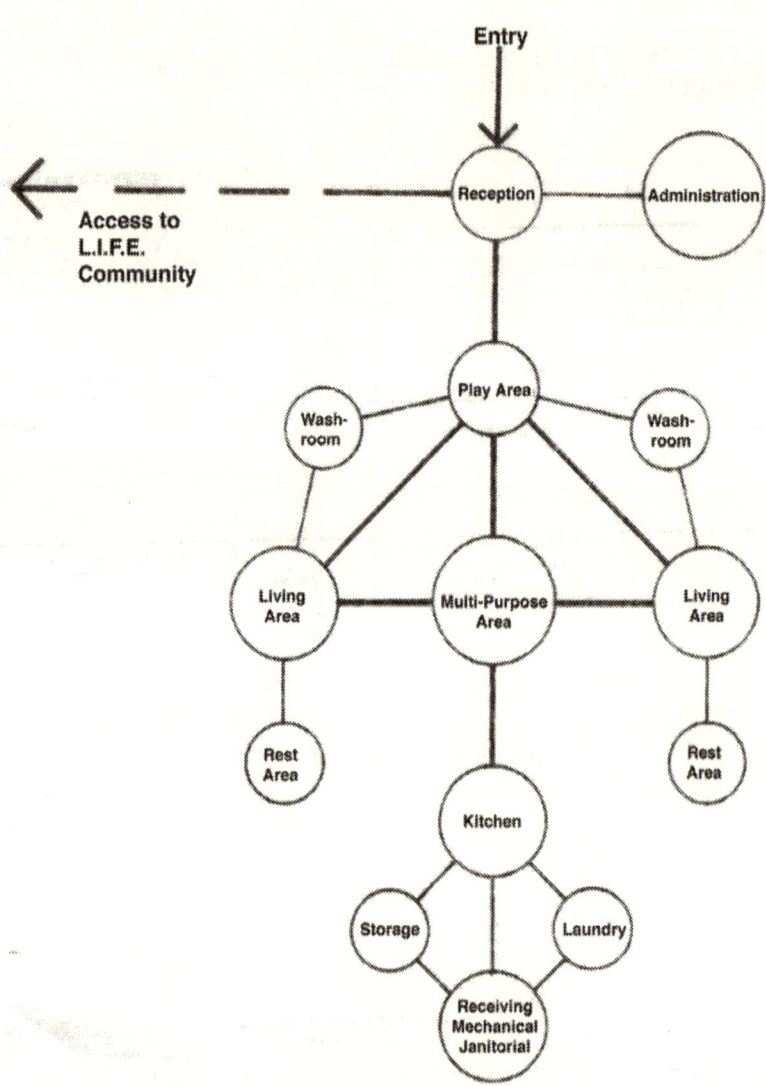

Urgent Care

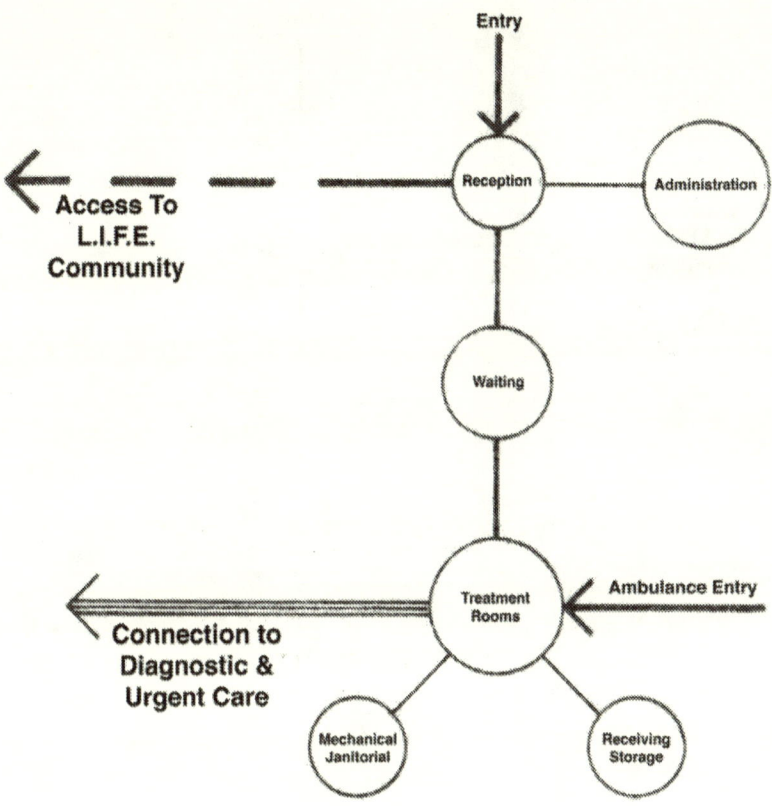

Education & Awareness

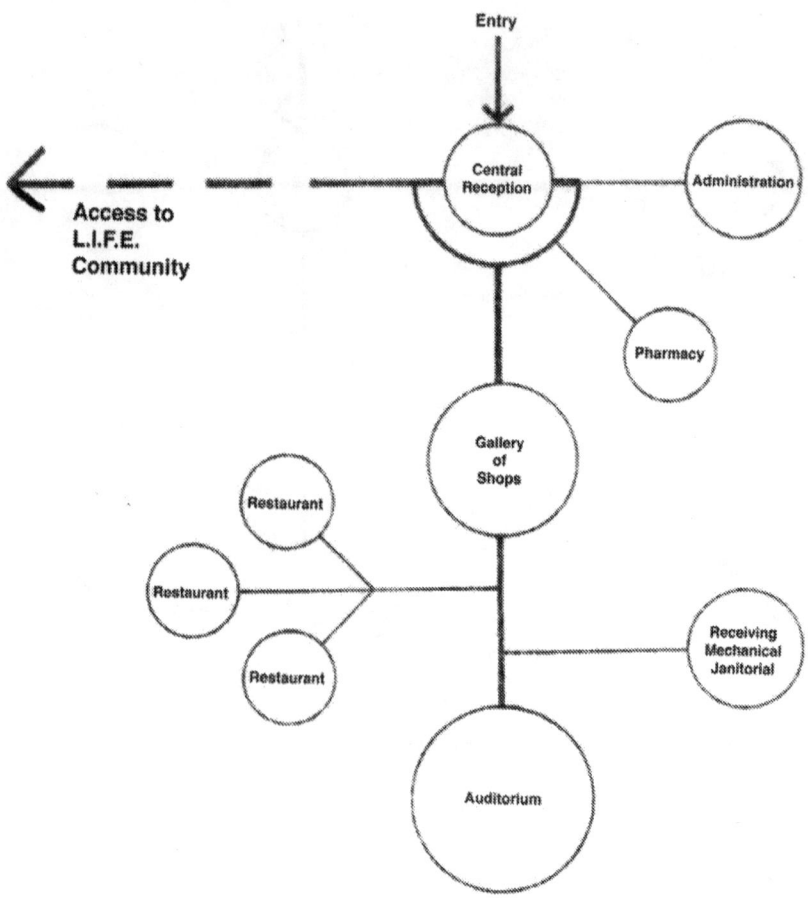

Each Community will combine those facilities best suited for them with the hope that balance will be the ultimate goal. Remember however, we are talking of many Communities throughout the nation and they will all share common, identifiable elements to establish a strong association, regardless of proximity. This allows each Community to be a visual marketing tool for both itself and the Life Network. One of the major objectives of the L.I.F.E. Community is to enhance Community solidarity. The full facility will provide a place where everyone, of all ages, can comfortably play and grow together. Health care suppliers and providers, who service the Life Network will enhance each others work as they join hands in making this National Community Project a reality.

Each Community will be geographically defined in a manner to pose no economic threat to another. Besides extended research into the broadened field of Wellness, each Life Community will provide an ideal setting for Medical Schools to train students and residents outside of the hospital environment. Research being focused on Health, rather than Illness, will be unique both for the students and their respective communities. Since all of this is exceedingly new, the facilitators of each community unit must possess the courage and foresight essential to the successful implementation of any major projects of social value.

Where Does the L.I.F.E. Community Fit Into
The Health Pyramed?

If one examines the L.I.F.E. Pyramed on page 258, G, H, and I represent the Economic base upon which the Pyramed rests. F, represents the L.I.F.E. Community and all of the facilities and programs that have to do with Wellness Medicine. E through A represent the forces of Traditional Medicine, and the Technical and Clinical services they have to offer.

Who Are The Private and Public Participants Who Must Become
Involved, for the Successful Implementation of This New System?

It will require the combined efforts of people and institutions from all areas of Society.

1) The L.I.F.E. Group Inc.
2) Health Care Providers.
3) Health Care Suppliers.
4) The Community.
5) Religious Organizations.
6) Business.
7) Industry.
8) State and Federal Government.

Wallace Salzman

Where does Federal Responsibility Lie
in Relation To The Health Field?

Up to now we have found fault with some of the decisions our government has made regarding Health Care. It is time to examine the good things that came, via the government that meant a great deal to the people of this nation and many others throughout the world. There are many who refute the argument that the medical profession is directly responsible for the increase in Life expectancy. They maintain that prolongation of Life is due to a decline in Infectious Diseases of the young and the prevention, via immunization, of the decimation of Diphtheria, Cholera, Typhoid, Small Pox, Dengue Fever, Whooping Cough, Scarlet Fever, Rheumatic Fever and Influenza like diseases.

Those who deny that the medical profession can take responsibility for the above good, claim that the decline was more due to the improvement of the environment and in the general resistance of the people who could then defend themselves more successfully against virulent germs. The data supporting this contention is also powerful. It is related to the institution of Public supported Mandates that redesigned the environment of the First World Nations.

1) Better Nutrition.
2) Purer water.
3) Better sewage disposal.
4) Better food hygiene.
5) Pasteurization.
6) Better jobs, pensions and social programs for the Elderly.
   These encouraged a decrease in the size of families and a decrease in the population explosion that occurred even before birth control pills.
7) Immunization against Small Pox, Tetanus, Diphtheria, Pertussis, Polio and Influenzal Meningitis, Hepatitis B etc.

If Society, with the help of government, can accomplish so much by having good Health goals and in this way integrate the talents of the community, it serves as an example of what the government and its people can do by integrating the Community with common goals and via this route, helping to integrate Man, the Individual, and via Man, the Community.

This is the goal of the lesser and greater L.I.F.E. Community. It is the brain function that lies at the base of Health and Illness. It is human thought and behavior that now designs both the functional and depressive Illnesses that remain at the forefront of our Medical Problems.

Sir William Osler once said "It is more important to know what sort of patient has the disease than to know what sort of disease or organism has the patient."

We assert that the designed environment of the L.I.F.E. Community and the knowledge and programs that will be implemented within its structure will become the next step in the evolving process of Health Care within our nation. It will become the prototype for changes that will one day be made available to the world.

The integrative mergence of a Pyramed of Services, with common values, common goals and common inter relational trust, along with a strong belief in the wisdom of both the vertical and horizontal hierarchy, should redirect our energies toward new frontiers in Medicine.

The Tasks of L.I.F.E.

We want our hearts to beat more slowly
Our lungs to breath more deeply
Our joints to move more freely
Our bodies to be stronger.

We want our heavy to be thinner
Our children to grow into healthy adults
Our adults to rediscover their healthy Inner child.

We want our anxious to find serenity
Our stressed to relax
Our sad to find joy
Our depressed to find goals, direction and excitement.

We want our minds to be clearer
Our spirits to be lighter
Our future to look brighter
Our LIVING to be longer."

Beth was through with her long presentation and waited for Jonathan's response. He appeared concerned and possibly a little confused and did not respond quickly to the period of silence that followed her presentation.

Chapter 39

Jonathan and Helen looked at Beth, startled at the amount of work she had put into her month of study. The rest of the class just smiled, all aware that Beth had added a whole new dimension to Dan's project and all because she had maintained a relationship with Nathan whose outstanding talents were still not completely understood. Jonathan turned to Dan.

"How does Nathan's concept fit into the Ortho-Para research center?"

Dan replied "It doesn't. It's too big for the property already selected, and to make it happen in the future it can't be entirely on one campus. I think that the concept is excellent, but putting it all in one place means creating Hospital-University campuses in areas of the country that could never afford it. Frankly Jonathan I think the Ortho-Para Institute, that Nathan wants dedicated to his grandfather Ben Yosif, is a good first step in testing out the merging of Alternative, Traditional and Ortho-Para Medicine on one campus.

While this project is being implemented, I will see how many acres of surrounding land is available along Long Island Sound so that in the future we can broaden our goals to include the dreams of the Life Group. I've heard of Syntec Corporation and I know the Architect in charge. It would be a wonderful addition to our project."

Jonathan looked at Dan strangely and finally asked "I'm curious Dan, isn't this friend of Nathan who suggested the Life Group, your father?"

The room suddenly became still as Dan restlessly changed his posture. He got up and took his place in front of the class with a growing smile.

"Jonathan, yes. Nathan's friend is my father Steve, who has proudly and happily kept his distance while I worked on the Ortho-Para project. He has kept his distance, while quietly watching me mature in my Architectural career and I know he would love to eventually become involved in a joint project with me that would fill the rest of both of our lives with challenges that would last eternally.

But Jonathan, I'm not ready for that immense jump. There are enough experimental mysteries that lie ahead with just the integrating of Medicine with Ortho-Para theory. When we begin the first project we'll have to create a team that can study the societal problems that will rise in relation to the Life Group Center."

Dan was finished and obviously surprised at what Beth had laid before the group. Jonathan, on the other hand, felt that the differences of opinion had to be aired if serious mistakes were to be prevented.

Jonathan entered the argument. "I assume that Beth presented these arguments of Nathan and Steve because she felt it was a feasible addition to the Research Center. But I agree with Dan. Even though the all inclusive notions of a Life Group Center sound magnificent, especially for the elderly, we have to first test the feasibility of the Ortho-Para test when it is merged with the concepts of Traditional Medicine and the various and sundry Alternative Cares. We have to see if effective space allocations are available for these yet untested therapies for those who are arrested in the different Ages of the Endophilic and Mesophilic Eras of life. Not only are the afflictions that cause illnesses of variable nature, but they effect those in different maturation phases differently. Several months ago while visiting Becky, Dan and I had a meeting with Dr. Su Shi and his son and went through the exercise of classifying the polarities of the different Alternative therapeutic techniques.

What we didn't do is classify the structural and functional Illnesses that require these Ortho or Para therapies and what's even more important, what preventative therapy we could successfully use to slow down or prevent an Illness, of either polar extreme, from surfacing. It is this latter professional skill that fits into the classification of Wellness Medicine and when wisely implemented would no longer require the services of an Illness physician, but preferably the wisdom of a Wellness practitioner who is well versed in all aspects of preventative Medicine.

We must remember that during the first year of life, when it is normal, natural and essential for Para dominance to control the egocentric behavior of the infant and child, that the inceptions and perceptions work together and control the unfolding, and active templating of the conceptual pathways, structures and synaptic junctions of the young brain. Let me ask Helen to speak on the slow perceptual and inceptual maturation process and in what way, by age of 12, we are witnessing a painful introduction to the ending of the Endophilic era and the beginning of the Mesophilic. Helen is the expert on the clinical aspects of Ortho-Para research."

"Thank you Jon. I'm glad you asked me because I recently enlarged my thinking on the concepts associated with the early programming of the human brain. Let me read you some of my recent notes on this maturation process.

'In our history of creation, from the nothing called Mattergy to the fabulous

287

conceptual weaver called the Brain, we have seen an evolving process that magically demonstrates that the more organized and integrated matter becomes, the more knowledge and wisdom becomes a part of the evolving process.

In describing the maturation of the infant during the Endophilic Era, we emphasized the unfolding of inceptual (feeling) awareness, the inner needs for the essential nutrients from the outside world, oxygen and food, and precipitating the breathing and eating processes, and the eventual excretory processes that were responding to inner bowel and bladder pressures.

Each specific organ of perception has inner values relating to life and survival. The perceptual organs respond to the size and character of the essential images that they are designed to evaluate, and also those that are dangerous to accept. In this way we can contract on vital and acceptable sources and expand away from unacceptable ones. The feelings (incepts) are the internal protective evaluators. The percepts are the outside signals that are to be evaluated. The perceptual organs are designed with the specific pick-up receptors sensitive to atoms, molecules or the ionic contents of the encompassing and surrounding environment.

Nose-

The olfactory system picks up odors or aromas made up of small molecules in flight. The aroma is labeled acceptable and desirable, or unpleasant and irritating, and the processes of moving towards or away from are in this way precipitated. An example of species essential smells are the pheromones created during a potential mating period.

Mouth and Tongue-

Lingual function and taste evaluates larger ions and molecules that are in solution, are partially or completely soluble, and represent some simple atoms or complex organic molecules that are either essential to life or potentially toxic. If the tongue fails to distinguish between that which is safe from that which is toxic, it would negatively effect other organs so that a toxic disaster could be in the making.

Eyes-

Visual function offers the differential between large complex energy radiations that are either essential or dangerous to the eyes or body. It affords us the privilege of either contracting on what we see, or expanding away from potential danger. It gives a stimulus to the functions of the pineal gland and its subcortical attachments that wake us up in the morning or, with the absence of light, allows the mechanisms of sleep to take over.

Ears-

Auditory sensations refer to the vibratory frequency of sound waves that we can either enjoy or find offensive. The experience of the listener determines that which they can appreciate and that which they would prefer to move away from. It is not only the volume, frequency, pitch and rhythm of the sound. It is also the linguistic content, the poetic rhythm and the musical content that may be species specific. The underlying Ortho-Para polarity of the listener also plays a role in either moving toward or away from.

The Skin-

The vibratory sense, intensity of touch, production of pleasure or pain, intensity of heat or cold or excessive or comfortable pressures are all the powerful monitors of the affect of the encompassing environment that is in close contact with our bodies. Protopathic pleasure allows and encourages close contact, whereas epicritic pain precipitates distancing. Protopathic sensations require Para dominance and Epicritic sensations require Ortho dominance.

As long as childhood Para dominance prevails, the dominant forces that influence behavior are:

1) The inceptions (the feelings).
2) The response to the feelings by the intrinsic central systemic environment. (The maternal-infant dyad).
3) The protopathic sensations of touch, taste and smell are the three most primitive perceptions.

The Endophilic, Egocentric values that dominate vision, hearing, and language exist in the child primarily to justify all of the above during Endophilia. When older, Reason, through experience, begins to challenge these sub-cortical processes that are present at birth and Ortho dominance begins to show its initial strengths as the concept of Allocentricity begins to show its value.

As Ortho grows in relationship to the systemic challenges related to contact with the father, siblings and friends, in and out of school, the unequivocal egocentric position is by necessity challenged by uncontrollable environmental systemic pressures and slowly the existence of other people begins to introduce Reason and a balancing of judgment.

By the age of 12+ enough failures related to this period of Endophilic egocentricity must take place before the awareness of the power of the self, to take care of oneself, begins to rise to behavioral dominance. Self sufficiency becomes a goal that challenges the concept of dependency that has dominated behavior since birth. Endomesophilia is beginning to blossom into existence and it must be re-enforced by the primary birth systems, mom, dad and family, so that the child is encouraged to further develop an Ego-Allo tension that further stimulates the processes of maturation.

Problems use vision, language and reading skills and a growing sense of power, to cope with the ever increasing concerns that rise as our perceptions of the outside world present them. As Ortho increases, because of the systemic challenges of the outside world, the inner inceptual values weaken as the dominant force that influences thought and behavior. This Para weakening represents the growing Ortho veil that is responding to the Ortho strengths that more effectively cope with inner systemic pressures. If Ortho dominance continues to be more effective in the competitive and sustenative outside world, it strengthens and becomes an effective guide to contend with most of the challenges of one's inside world but hopefully not with the intimate world of marriage and family.

The greater the number of outside systems, the more intense the Ortho polarity, and the more difficult it is to revert to Para and the behavior that is essential with the more intimate central systems. It is this conflict between Para and Ortho behavior that produces the inner conflicts that become the functional illnesses with which everyone must contend with as we travel on the road to maturation.

This is the period of Mesophilia where power soon grows to its greatest level and in most people fails to discover the compromises that must be made with the intimate central systems that demand the Para, and the more dependent awarenesses, associated with the passionate areas of life. The passionate central system of marriage and family are an extension of the maternal-infant dyad, where the essential dependencies can be accepted and worked on, even as the drive for independence takes on the challenges of the outside world.

In Endophilia, we felt dependency and wanted the external world to revolve around our personal needs and feelings. Our inceptions dominated our growing mind.

In Mesophilia, we earned our independency and wanted the power to control our outside world, so frequently forgetting that our intimate world is not outside. The love that was given to the vulnerable and dependent child, that was so effective and essential in our life with mother, was replaced by power and demands which are universally ineffective in the world of intimacy. Our perceptions of how the outside world works dominated our behavior.

With the birth of some Ectophilic strengths comes the birth of balanced concepts regarding a multi-systemic life. The notion of seduction and the dependency of Endophilia, and that of power and the total independency that relates to Mesophilia, slowly finds its balance in early Ectophilia, as the potential for the creation of the Embryonal Chameleon, we've described on so many occasions, comes into existence. This is the condition required for the balance between Passion and Reason, Para and Ortho, as we act out our roles in the complicated, multi-systemic dramas of life."

Chapter 40

Jonathan, once again, took the floor in utter amazement at what had occurred during this long and complicated discussion regarding what turned out to be a far more complex answer to what Dan and he had anticipated. The class was looking beyond both of them and it became evident that the answer to the questions regarding the architectural design of the Research Center would have to take into consideration the Medicine of tomorrow. Jon and Dan would have to make phasic decisions, aware that the Medicine of tomorrow very quickly would not be the Medicine of today.

Wellness Medicine, which is a minor system now, and which is loosely attached to the major concepts of Illness Medicine has influenced the trained Allopathic physicians and Alternative Care Healers of today. It will someday change and this huge behemoth has many problems that are already being challenged by the scientists in all major University centers throughout the world. Ortho-Para theory is the base for some challenging suggestions regarding the practical management of disease in general, but the field of genetic research and to a degree, many of the behavioral discoveries regarding human cellular structure and function are going to one day open new fields of research that will broaden the challenges of the trained physician even more.

Jonathan became aware of a new challenge to Medicine he had not anticipated so soon. As fast as Medicine was changing, his concepts, and certainly Dan's would have to become more flexible, or the decisions made today could create new and expensive structural problems that would be difficult to remedy.

All of these thoughts were going through Jonathan's mind as he faced the group. "This has been a long and arduous meeting. It has been a wake-up call to the problems of tomorrow that we must anticipate as we design a structure flexible enough to one day encompass the Medicine of Tomorrow.

Ortho-Para, V

It made me very aware that the Ortho diseases and Para illnesses have to be more clearly defined so that they are more easily differentiated than they are today. During the initial phases, differentiation will hopefully become easier when we get more aware of all of the disease etiologies, but at times the Para illnesses, which are at first functional and the Ortho diseases, which begin as structural problems, become mixed diseases and require combined therapeutic modalities that encompass the entire panoply of Ortho and Para therapeutic approaches to the problems we initially might have solved more easily.

With this in mind, I'm going to ask Kenneth, Allison and Cassy to develop a preliminary Index of Differential Diagnosis which is different from those voluminous indices available today. They should be divided into the major categories we defined previously. I realize that at times it's difficult to know what came first, the functional or structural abnormality, but this is our first try and let's see whether a fine line can be drawn between these classifications, and if not, we'll then decide how we can remedy the problems that arise.

Dan, on a basis of what you've learned and noted down since this meeting began, I want a rough plan presented to me in about two weeks. Take into account that your present assignment is only the beginning of an eternally growing project that must have no end we can foresee.

While you're all pursuing your goals I'll be doing the same thing. If anyone in this room can assist Dan in resolving the problems presented today, I'll be grateful. I have in mind that Dr. Su can help all of us with his practical knowledge of Chinese Medicine and possibly the methods of getting strong, reliable Herbal Medications.

We are all exhausted, so I'll bring this magnificent meeting to a close and wish you all good luck in your creative work."

Jonathan was overwhelmed by what he had heard and was ready to rest and incubate all of the seeds of wisdom that unexpectedly had come his way.

Jon and Helen, with heavy heads and enlightened minds, lay in bed that night quietly reviewing the mental challenges and images of that day. They had no inkling of what the complexity of that day's meeting would bring from these multi-talented professionals who were seeded with multi-goaled notions by Jonathan, only 25 years before.

Helen turned to Jon "Do you realize what this class of yours has accomplished? They have digested and assimilated your theories and just as you inoculated them with the basal knowledge of the Universe, they have taken it and run. The tree of Ortho-Para and Systems and Particles now have so many branches, it is, to me, a miracle."

She turned on her side and looked at Jon who was smiling with quiet joy. He took his hand and touched her naked breast. "Yes, I know darling. That class of youngsters proved, without question, that if you create a questioning curios-

293

ity in young ones there is no way of measuring how enormous can be their growth as they mature.

Dan already has a rough drawing of the Research Center and I think with a little polishing, he'll have structural plans ready to show me in possibly two or three weeks. Nathan has already found another 60 million dollars for the project and is amazed how little resistance he's meeting from the Foundations to whom he's given only a rough estimate of what we're trying to accomplish.

Ed Johnson of the Johnsonian Foundation knows about our project and he called me and wants to invest some of his capital from the Institute in our Research Institute and even encourages us to go big and incorporate the Life Concept that would set the stage for a radical change in Clinical Medicine in this country. I told him that such an undertaking might cost one half to 1 billion dollars. His answer was 'The money should not be a concern, I know it's there'. Then he wished me good luck and assured me that any basic science that comes out of the Johnsonian Foundation will be ours for the asking. It's marvelous. Suddenly our growing pains have been worth the outcome they produced and the magical people and friends we met along the way."

Helen was listening quietly as she was sprawled out, nude and uncovered on the bed in what seemed like a reverie as she responded to Jon's poetic rhythm, as he was recapitulating the few things that had happened in the past few weeks. She responded "I know that everything is suddenly coming together and I think it's time for us to do the same. It's been nearly two weeks and your touch is mesmerizing me."

Jon didn't need a second hint. He responded to the go ahead signal with youthful enthusiasm. Sleep came with sheepish smiles only shortly after.

Chapter 41

At breakfast, the next morning, Jon was very quiet and in deep contemplation. Helen went about her chores and did not disturb his train of thought. She had learned not to disturb him when he was trying to make, what appeared to her, a serious decision.

This went on throughout breakfast in a very quiet house, without any interchange between both of them until it appeared to Helen that the change in his facial expressions indicated a decision had been reached. He looked up from his cereal and turned toward her.

"You know darling, I was thinking that between Kenneth, Laura and Daniel, there is nothing that passes through my scatter-brain mind that is foreign to them. Whether it has to do with the principles of human behavior, evolvement and maturation, growth and knowledge, they are on a par with us and are ready to take off for where? Only God knows." He smiled as he got up from his chair and approached her.

"You know, I don't know what else I can teach them. After yesterday, I admit they have a load to teach us. It is as if I've had 17 children and now with their marriages, it has grown to 34 plus. I think it's time for you, dear Helen, to take center stage. Larry has already won the Noble prize for his PET scan work on the Ortho-Para nuclei of the brain. I will be watching the Ortho-Para tree grow in its theoretical powers and I think it's time for you to make your Ortho-Para test available to those physicians who've taken your course in Ortho-Para Medicine. It would be wonderful to have a large entourage of knowledgeable young practitioners ready to take over positions in our Institute when it opens 2-3 years from now. Are you ready to go public?"

Helen smiled and walked over and kissed him while he appeared to still be in a slight trance. "Jon, I was ready a few years ago but wanted to train a few more doctors before we went public. I am now ready and I'm sure that in 2-3 years we will easily be able to staff our Research Institute. Ben Yosif would have been pleased and I'm sure Nathan will be thrilled. But darling, you're only 55 years old and you're talking as if your creative eurekas have come to an end. I would guess that you have 10 or more years before you peak. Certainly Ben Yosif didn't admit to peaking until he was in his mid seventies."

Jon smiled and kissed her hair. "I know that dear and I still have a lot of questions that plague me, but they're more involved with Dan's problems and I'm more focused on all of his problems than I am on the behavioral abnormalities of mankind. It's sort of an ideological hiatus. When his problem is solved, I'll again be able to focus on the behavioral Illnesses of human interaction. I'm not really worried darling, so please don't you worry."

Helen rose from her chair and walked toward the sink. "You know Jon, I'm not worried, but you sounded as if your work had come to an end and in truth you're only at the starting gate, just before the race begins. These past few months were necessary to solve a load of functional problems. Laura and I will handle the problems related to the Ortho-Para test and Script."

Jon seemed pleased with Helen's decision. "That's wonderful dear. I'm going to Dan's office today so that I can be available to immediately respond to any questions and to work on a new idea that was seriously suggested by Ed Johnson. He felt that the money necessary to go ahead with the big plan presented to Beth by Nathan, who was under the influence of Dan's father, was an achievable dream. Ed felt that the half to one billion dollars to make it happen was not unrealistic. I know that Dan appeared somewhat frightened by the immensity of the goal and honestly, I was too, considering the cost. But I think I'll quietly study this possibility without suggesting to Dan, otherwise. The 'Ben Yosif Ortho-Para Research Center' can be our first project, with solely the goal of Integrated Medicine involved, especially because of the limitation of the lot size. It would teach us a great deal about the possible problems of Integration, so that we can avoid them, if someday in the future, our son Ben is ready to take over a project which we're too old to complete and oversee and we'll just have a perpetual smile on our faces while our son takes over the control of the babies we created."

Helen began laughing hysterically. She looked at Jon with wonder in her eyes. "Jon, stop talking that way. You're talking like a man who thinks he might conceivably have control over his son and his professional decisions regarding the future. Remember dear he has both your and my genes, and if you recall, neither of us had parents who were unwise enough to try to influence the direction we would be going. And that was good. We were forced to broaden our thinking and allow our own personal eurekas to play a role in the ultimate direction we would be going. Ben is way ahead of both of us and has yet to show any interest in what we're doing. Let's keep it that way."

Jon joined Helen in laughing and apologetically responded, "Darling, You are absolutely right and I won't deprive him of the magic of his own creativity by making him feel a responsibility for the future of the Ortho-Para theory. That's Ken, Ally and Laura's responsibility. I, for a moment, reverted to the powerful egocentricity of my youth. That won't happen again. I'm leaving now for Dan and I'll see you this evening." They kissed and parted, Jon sheepishly smiling because of his foolish comments.

The next three weeks went by very quickly. Dan and Jon were going into conference 9 or 10 times during a work day, as the gross sketches of the Institute began slowly to unfold.

At home, Jon was initially quiet about his Alternative decision but he suddenly became talkative and Helen looked with curiosity at Jon who finally decided what therapeutic Alternative modalities would be chosen for the new Ortho-Para Research Center. With much curiosity she spoke, "Well Jon, what have you decided?"

Jon hesitated only a moment and then with a big smile he responded.

"Darling, you know that I have to be careful. There is no way that I can cover over 100 of these Touch Therapies and I must trust my intuitive sense. The new hospital will have all the Ortho invasive therapies that exist in all the other University hospitals. With this in mind, I've chosen the following Alternative Therapies:

Yoga- Yin and Yang
T'ai Chi
Aikido
Traditional Chinese Medicine
Homeopathy
Aroma Therapy
Reflexology
Chiropractic Therapy
Massage
Light Therapy
Autogenic Training
Hypnotherapy
Magnetic Therapy
Psychoneuroimmunology
Divine Healing
Chanting
Body Works
Spiritual Healing

Chapter 42

It was now Helen's turn. Although her Ortho-Para test had been written about in the Journal of Behavior, the test itself had never been publicly made available to the Medical profession and if it was going to act as the indicator to whether Para or Ortho therapy should be recommended, both the test (OPT) and the method of interpretation (SCRIPT) would have to be made public.

Jonathan had, for years now, been concerned about making it available to all physicians, but there was a serious problem. Not enough doctors had been trained in its use, and its relevancy could only be based on a large number of patients who had been evaluated and treated according to its findings.

The time was now ripe. The Medical Criteria for taking the test and the method of evaluating it was now complete. Patents were now in force to protect Jon and Helen from a plagiaristic infringement on their rights, and the legalese regarding their protection was now in place.

There was yet one unsolved problem regarding the publishing of her work. Should the test be available to the public, at large, or just to the physicians who would use it in their evaluation process? Was the SCRIPT analysis written in such way that the patients would understand the recommendations? As Helen re-approached these problems she suddenly remembered that one of Jon's prime concerns was in changing the relationship between physician and patient. It was no longer to be rescuer and victim relationship, but teacher and student. The test and its interpretations had to be made available to the patient and if it was their desire to seek out a consultant, the patient could then seek out an Ortho-Para consultant who would clarify the meaning to the answers in the Ortho-Para test. It was an opening wedge to changing the relationship between doctor and patient, requiring the physician to be involved with all of the patient's Systemic involvements, as are third world Shamans.

The answer to her dilemma was obvious. The test and the analysis had to be available to the requesting Public and the chore of the physician was to be available as a knowledgeable consultant when the patient chose to understand the dynamics of his/her problem and expand beyond the recommendations of SCRIPT. Having solved that problem, the only thing that had to be done was to get in touch with her publisher.

(3 months later)

Her decision was correct. The OPT and SCRIPT were only on the market for one week when to Helen's astonishment, they were in such demand, the publishers were forced to sub-contract the orders to other printers in order to meet the immediate demand. Helen and Jon were elated initially, but with its growing success they discovered that the number of doctors capable of helping those who wanted help with Script interpretation, were too few and the demands by M.D.'s and Healers was so great that Helen had to rent Conference space at the Johnsonian Institute to teach Ortho-Para theory and OPT interpretation to physicians who found the demands of their practice overwhelming.

Even though the Ortho-Para concept had been offered via a post-graduate course to physicians for ten years, it took a demand by the patients, to wake up the practitioners of both Allopathic and Alternative therapies before there were any responsive reactions by either the educators in Medicine or their students, the Practitioners. The practice of Medicine was being returned to the people and apparently it had to go through a labored birth and serious crisis before it slowly was accepted by the rigid skeptics, the professionals who were prone to change only very slowly when there was intense pressure put upon them by their patients.

In the three months that passed, Dan had finally completed the drawings of the primary three buildings in which Ortho and Para Medicine would be further investigated and which included the two buildings in which Ortho and Para Practitioners would be able to practice on a basis of the OPT and SCRIPT analysis by Helen and her Research group.

A Summary of Life's Autonomic Journey

We begin in Para
- The dark caverns of complete dependency lie in the primitive sea of the protective amnion.
- The amnion incorporates the embryo which is responding to 3 billion years of ontogenetic memories, as they recapitulate the origin of Life.
- All creatures and Man.

Feelings have yet to awaken.
- As one phylum turns into another, the embryo floats in the Love waters of Peace and Tranquility.
- One day this will be the goal of embracing lovers, who are heatedly fused in conjugal embrace.
- See them responding to the oscillating rhythms of their choreographed bio-dance, as they once again create a moment of reversion and return to the uterine love haven where life first began.

We are not finished.
- Our journey is yet incomplete. We must travel from this primitive world of Para through a cervical door along the dark rugated passage of woman, and plunge into the world of smells, taste, touch, pain, noise, and light.
- Our nose, tongue, ears, and eyes have been awakened to another world and it is now the world of Ortho and Para.
- Chaos, and confusion has been born.
- The infant senses terrible needs, but doesn't understand them. Suddenly a cough, a scream and the lungs begin to move. Inner turmoil has initiated the comfort of breathing. Infantile chaos repeats again and again as the visceral signals of undifferentiated and fractional instincts are given meaning by feeding. The outside world gives understanding to the inner negativity that will only disappear when Para instincts allow the Ortho outside world to serve it.
- The tussle between Ortho and Para has now begun.
- In the new world of Ortho we find sustenance, protection, and energy. In the ancient world of Para we find regeneration and potential.
- During our infancy, mother's breast is the haven of Para.
- During childhood, our father's demands are the symbols of early Ortho. During adolescence, our siblings become the source of Ortho strife. During pubescence our teacher's demand our virtual Ortho independence.
- During young adulthood, work and rest create the balance between Ortho breakdown and Para reconstruction. During middle age, we have reached the peak of our potential. During old age, we are slowly approaching the time of reprocessing.

It is at this time that the bonds of Para begin to uncouple.
- First we retire and leave our work
- Then we leave our friends
- This is followed by detachment from family
- And then finally from Life.
- Our Mega-molecules become Monomers and they in turn become the elemental base of structure.

From dust we have become Man.
From Man we have become dust.
From Para we have turned to Ortho.
Our Autonomic life is now complete.

Lexicon of Words

A

Adrenergic- pertaining to those chemicals, similar to adrenalin, which charge the body with high energy and stimulate the Orthosympathetic (Ortho) Nervous System and its dominance over the Parasympathetic, its antithesis The Catecholamines of the brain that awaken the arousal centers of consciousness cause this Ortho dominance. An alert mind, well toned muscles, a pounding heart are the result of this dominance. If in excess and out of balance with the Parasympathetic, it leads to anxiety, manic states and even to convulsions.

Affect-a general term for all the signals generated within the body because of shifting Autonomic dominance throughout the body because of the perceptual signals known as touch, pain, selective warmth, sight, hearing, and taste. The Perceptual organs of the body selectively screen relevant visual, auditory, olfactory, gustatory, tactile and genital signals. Thus affect covers the broad concept of feelings and Perceptions and represent the signals that precipitates emotions, posture, and behavior.

Afferent- an anatomic vectoral term representing those nerve pathways that bring outside information into the body. The Perceptions travel on afferent pathways as they bring signals of specific type to the brain for storage, analysis and action. It is in contrast to the efferent pathways that travel to all of the muscles, and whose signals cause contraction of body tissues. All movements, voluntary and involuntary, are due to efferent impulses.

Agape- a unique form of Love that is totally selfless in its priorities. It is the love that gives without receiving, that loves without expecting love, that is generous with anonymity. To help someone with expectations of reward in another life is a subtle contamination of Agape which seeks no reward.

Allopathic- The physician of today who treats the sick.

Anthroprogenesis- a multisyllabic word that refers to the beginning moments of Man. To the Evolutionist there is still difficulty in defining that moment when ape became Man or some Anthropoid like man grew a skull with a large brain capacity, large enough to fulfill the anatomic qualifications of the " I think, therefore I am" kind of man.

Antidromic-is a neurophysiologic term referring to an afferent nerve that is acting like an efferent. Retrodromic is a synonym for a nerve that is carrying a current in the wrong direction.

Akashic Record- Comes from a Sanscrit word which means (Primary substance) or (that out of which all things form) They are the imperishable records of life from which all things formed and this only when man's thoughts are in sympathetic vibration with its Akashic origins. It is the supreme intelligence that a totally deterministic and Preformationist concept of Life demands.

Antitheses- Opposites. In our writings, it has been a growing concept and while explaining the idea of balanced opposition has taken on the power of a Universal Law. In my late years, my very goals were moderated by an awareness that no thesis existed without its antithesis.

Allocentric- (Other centered). The opposite of Egocentric.

Actualization- It encompasses the notion of human growth, potential and maturation.

Autonomy- A relative term for independence.

Anaclinic- Leaning against or dependent on. If mother leaves the babe it can die of anaclinic depression.

Anelectrotonic- A condition of decreased nerve sensitivity in the region of a positive electrode or anode.

Askesis- An Indian term described in the Gita. It refers to rigorous training, self disciplin; as in an ascetic.

Asuric- An Indian classification of those in India who do not have the ability to climb out of the more feral state of childhood.

Avatarhood- It is the incarnation of God and the Indian equivalent of Christ.

Anabolism- The structure building process of human metabolism.

Authenticity- A term coined by Sartre. To live life in relation to that which molded us.

Ayurvedic- Medicine that is practiced in India.

B

Bathos- Comedy of Life; humor.

Biogenesis- Life- Recapitulates the evolution experience of all matter

Bimodal Consciousness- The control method of the right and left Brain working together.

Bipolarity- Refers to the concept of opposites and their attractile or contractile potential. Ex: pos-neg, male-female, north-south, are all polar differentials.

Bisociations:- When two conceptual notions cross at a point of commonality, it is the point that a new notion is born.

Buffering- By which the constancy of the environment is not changed. It controls the constancy of the pH.

C

Casuistry- The study of doctrines of moral absolutes dealing with conscience and its specious conclusions.

Catelectrotonic- The local depolarization and increased irritability in the region of the negative electrode or cathode.

Cathexis- A name used be Freud to describe love-object attachments.Cybernetics- The science of feedback mechanisms.

Chamelionism- Refers to responsive skin changes when exposed to different environments.

Chi- The Chinese term for all kinds of energy.

Chromaffin Cells- The cells that become the Ortho nervous system.

Coenesthetic- Synonymous with Global or Wholistic.

Cosmogenesis- Step by step creation of the Universe.

Cyclothymic- Another name for Manic Depression.

D

Daivic- An Indian term that speaks to the fact that with the caste system some people are prone to a higher degree of maturation than the Asuric.

Depolarization- The process of obliterating the distinction between opposite poles or antithesis.

Diencephalic clock- The theoretic and very quiet location of the biologic pacemaker controlling the endocrine hormones, their secretory time, and unconditioned life reflexes.

Determinism- States that Man has little control over his destiny.

Differentiation- The analytic process of separating a Whole into its parts.

Dishabituation- The elimination of a response to a stimuli caused by repetition.

E

Epistomology theory- States that the study of methods and bounds is a pursuit only of Man.

Epicritic- Discrete, pin-point, sensitive sensation and perception.

Embryonal Levels of Maturation- Endophilic, Mesoendophilic, Ectoendophilic, Endomesophilic, Mesophilic, Ectomesophilic, Endoectophilic, Mesoectophilic, Ectophilic.

Ectothermic reaction- Rrefers to those that create and give off heat.

Endothermic reactions- Refers to those that require the addition of heat.

Entropy- A rule of thermodynamics that all systems tend to break down into lesser and lesser states of organization.

Existentialism- A multiordinal term that fails to lift man above his least potential.

Exteroception- Those perceptions picked up by the skin.

Extropy- Coined by author when examining the process of functional antitheses. It is a system that causes an increase in the rebuilding of potential energy.

Epigeneticist- One who believes that along with genetics, the living organism responds to environmental pressures.

Extimacy- The opposite of intimacy.

F

Functional- A term that relates to both process and movement. It involves the use of kinetic energy.

Facilitation- The process of enhancement.

Fantasy- An unrealistic concept used as a base for normalcy of behavior and human expectations.

Field Forces- Defines those forces of contraction and expansion that govern the oscillation of every system of particles.

Foreordained- Balzac's term for the deterministic nature of cuckolding.

Fission- The nuclear term for atomic disruption.

Fusion- The nuclear term for atomic build up.

G

Generative- Covers the broad concept of creativity and birth. It is in contrast to the sustenative functions of life.

Graviton- The boson of the gravitational force field.

Gestalt- A subtle term that reflects on the mystery of summation. It reflects on a hunger for a repeat of a perception that has become part of habit.

Gunnas- The three major phases of maturation as described in the Gita. They are The Tamasic, The Rajasic, and The Sattwic.

Gnostic- One who has strong beliefs in the process of gnosis, a process of mystical intuitive recognition that equates matter with evil. Only the Spiritual have the highest good.

Global- Synonym of Coenesthetic and Wholism.

H

Homeostasis- A biologic term referring to the body's tendency to seek out specific levels of chemical equilibrium.

Hyperventilation- Primary and secondary over breathing to control a low pH or anxiety.

Habituation- An increase in tolerance to a drug so that more has to be taken to produce the same effect.

Hypothalamus- It is the subcerebral organ that controls the Autonomic nervous system.

I

Intrusive- An aggressive move into a system whether wanted or not.

Incorporative- The encompassing of that which is surrounded. It is in contrast to intrusiveness.

Intuitive- The creative surfacing of answers without conscious use of reason.

Identity- Is exact sameness in structure and function.

Immanent- That thought that lies within. Thus to Man, God is immanent.

Implosion- A sudden coming together such as fusion.

Inceptual- A new term to incorporate all endogenous pacemakers that through deprivation forces us to eat, see, and hear.

Inhibition- Essential prohibition of a previous allowed process.

Intrajection- The process of directing causes inward, such as blaming oneself for everything.

Involution- Opposite of evolution, such as reverting to a child after having demonstrated evidence of adult maturation.

Imprinting- A term that recognizes the critical phases of maturation and the critical effect of the environment during this phase

Isomorphic- A signal or sign that completely describes an idea that is understandable.

Instinct- An integrated sequence of automatic behavior that is an integration of an inner need with an outer source of its assuagement. It can misguided because of parental mistakes during the preverbal period of life.

K

Kabbalah- The cult of Hebrew Mysticism. It represents the Hebrew inner search for answers to Universal questions and is just as filled with fantasies as other mystic orders.

Kinesthesia- The inner perception that is automatically fed back to the Cerebellum by proprioceptive nerves. It automatically governs equilibrium and balance at a subconscious and basal level.

L

Leftist- Someone with Socialistic, Communistic or very Liberal tendencies.

Limbic- Pertaining to the Limbus and Rhinencephalon, two primitive parts of the brain that deal with feelings and emotions. Dangerous damage to this part of the brain causes major changes in affect and effect.

Latency Period- A part of sequential growth (6-12 years of age) in which sexuality is subdued and is transferred into energic industry. It prepares the individual for that time when sexual hungers will grow again and requires a competitive antidote.

Labyrinth- A maze that when entered creates directional confusion.

Logos- The conceptual creation of Philo on the island of Samos 4th century B.C. It says that reason alone is the truth and remains the only manifestation of God.

M

Mesoderm- The middle stage of the Embryonic first stage of differentiation.

Metabolic- A term that refers to the dynamic chemistry of the body via the route of anabolism and catabolism.

Mesomorph- A person who is built to accomplish feats of strength.

Mucocutaneous- The junction that represents the structural site of greatest propathic sensation. The mouth, anus, nose and genitalia. They are the sphincter orifices that allow bounded invasion of the body with highly sensitive organs of perception, sensing whether an invasion is good or bad, pleasurable or unpleasure. If pleasure and its goals grow in the sphincter areas, they can radiate their perceptual reversion to other areas of the body and by this experience lessen the critical epicritic pain.

Mattergy- The name of the undifferentiated energy that, at the beginning of the universe, turned into matter and energy. It is the name given by the author in his first book.

Minataurize- Balzak's word for cuckolding.

Mazdaism- The primitive base of Sun Worshipping upon which Zoroastriaism was practiced in Persia.

Metaphysics- The science of thinking that deals with non logical, non analytic inductive thinking which doesn't stand up to deductive proofs.

N

Nirvana- Pure Spirit. It is a reversion to the concept of transcendency, immanence, and the patriarchal concept of Supreme power in its primitive kinesis.

Neurobiotaxis- Refers to the functional and structural changes in the synapses in response to functional stimulation of that part of the brain.

Neocortex- That part of the brain considered by most as the greatest part of human evolvement. It encompasses the cerebral cortex.

O

Ontogeny- The process of individual evolvement through the process of Phylogeny.

Ortho- One of the most significant words and concepts in my writings. It is a shortening of the Orthosympathetic nervous system and refers to the energy mechanisms, best known by the effects of adrenalin. It is the antithesis of Para, the other part of the autonomic system known as the Parasympathetic system. Its function relates to structure building at night instead of energy building during the day.

Orthokinesis- The dynamic action when one is in Ortho dominance during a day of well motivated action.

P

Predeterminism- The belief that all of man's actions, goals and motivation were predestined by genetic inheritance.

Pacemaker- Cells that control the rate of action of an organ.

Pantheism- The philosophical and religious concept that sees God and Spirit in all that lives.

Para- See Ortho.

Particle- A structural entity that in integrated action with another particle makes up a system. Discussed in great detail in the first Treatise at the end of Ortho-Para I.

Partistic- Pertaining to the notion of separation and differentiation. It is the analytic method of solving scientific and life oriented problems.

Phylogeny- Is to species evolvement, what ontogeny is to individual evolvement.

Preformationists- See predeterminism.

Projection- The psychologic process of always putting blame for mistakes outside of oneself.

Protopathic- Pertains to a primitive mode of perception where discrete pin point sensations are lost in the integrated fusion of all. With damage to the sensory nerves, it is the first sensation that returns on healing. It may be the initial way a baby experiences the world.

Paleocortical- See limbic

Potentiokinesis- Is a general term of energy integration reflecting the common origin of both potential energy and kinetic energy and the ability to translate one into the other.

Psycho-cybernetics- Pertains to the continuous and moderating perceptual feedback that alters action and human emotions on a moment to moment basis.

Q

Quark- A theoretical particle, yet to be physically discovere but postulated to be the unit that turns into nucleons.

Quantum- An energy state unique to each particle and which is dependent on the orbital position within its structure.

R

Receptors- An area specific for receiving a signal of a hormone stimulation from another chemical called a ligand.

Refection- A biocybernetic term referring to the feedback characteristics of a cybernetic activity. When positive, it means that the reaction is stimulating its own activity. When negative, the activity is inhibited. When we have high energy and perform anxious tasks, there is a positive continuous growth of this energy and anxiety. This is the meaning of Ortho dominance that is excessive.

Reticular Excitation- Facilitation or inhibition of a network of cells in the Central Nervous system that have to do with arousal or awakening.

Refractory- A term used for a non-responsive system.

Resonance- The enhancing process noted when sound or E.M.W.s have the same frequency of vibration and merge. It is this process that enhances the frequency of Lasers and Masers.

Rajasic- An Indian name for the Guna that holds the men who aggressively attack life, and achieve by living a very physical route.

Rugated- The repeated wave-like structure of the vagina.

S

Script- It is the term used by Eric Byrnes to explain the concept of Cerebral tapes which precipitate all kinds of good and bad behavior.

Servomechanics- The patristic mechanics associated with Cybernetics and the complicated tools that work with constant feedback from the environment in which they function.

Super-Ego- The social antithesis of the unsocial Id. It is a Freudian term that explains the way that civilization and society attempts to mold properness into the society.

Schemata- The conceptual scaffold about which we integrate concepts of knowledge.

Syndrome- A group of symptoms that always appear together.

Sustentative- The actions performed to sustain Life.

Sattwic- It is another Indian term referring to those who have reached the highest position of attainment. It refers to the most religious in terms of the goals of Hinduism.

T

Tannaic- Pertaining to the Rabbis of Palestine during the first two centuries. They apparently wrote the Mishnah.

Tamasic- The lowest cast of India, considered incapable of living beyond the egocentric seductive phase of life.

Telos- Goal or ultimate aim.

Transactions- It is the give and take of mutual communication.

Transmutation- The changing of one thing into another.

Totemistic- The family that uses a Totem or family symbol to represent it in public.

Transcendental- Rising above and looking down upon. It usually relates to the concept of God.

V

Visceral- Pertaining to the organs within the body.

W

Wholistic- The global or coenesthetic type of diffuse perception.

Within-ness- A term used to describe the goals of intimacy and privacy.

Bibliography

Afzelius, Bjorn (1964)

The Anatomy of the Cell
Translated by Birgit Satir
University Of Chicago Press
Chicago, London

Albers, Bruce et al. (1969)

The Molecular Biology
of The Cell
Garland Publishing Inc
New York, New York

Appenzeller, Otto (1986)

Clinical Autonomic Failure
Elsevier, New York, Amsterdam

Arendt, Hannah (1958)

The Human Condition
University of Chicago Press

Asimov, Isaac (1968)

The Universe-
From Flat Earth to Quasar
Avon Books

Aurobindo, Sri (1974)

The Gita
Edited by Shyam Sunder Jhunjhunwala
Auropublications

Aurobindo, Sri (1974)

The Future Evolution of Man
The Divine Life Upon Earth
Theosophical Publishing House

Aurobindo, Sri (1962)

The Life Divine
A Quest Book
compiled by P.B.Saint Hilaire
Theosophical Publishing House

Baker, David

Disease, Pain and Sacrifice (1968)
University of Chicago Press
Chicago, Ill

Wallace Salzman

Balch James M.D.and Phylis C.N.C   Nutritional Health
                                                     Avery Publisher

Bichat, Xavier (1977)   Life and Death
                                                    (written in 1796)
                                         Translated by F. Gold in 1827
                                                 Arno Press, New York

Bonnor, William (1964)   The Mystery of
                                                 The Expanding Universe
                                         Macmillon Company, New York

Bradford, Nicki   Editor of
                                         The Hamlyn Encyclopedia of
                                                   Alternative Health

Buber, Martin (1978)   Between Man and Man
                                       Macmillon Publishing Company Inc.

Buber, Martin (1953)   I and Thou
                                              Edinburgh: T & T Clark
                                      Translated by Ronald Gregor Smith

Cassirer, Ernst (1953)   The Philosophy of Symbolic Forms
                                                 Yale University Press

Chardin, Pierre Teilhard de Chardin (1973)   Building the Earth
                                                         Discus Books
                                                     published by Avon

Chardin, Pierre Teilhard de Chardin (1959)   The Phenomenon
                                                              of Man
                                            Harper and Row, publisher

Cleckley, Hervery (1950)   The Mask of Sanity
                                                  C.V. Mosby Company
                                                 Biology Today (1972)
                                                        C.R.M.Books

Davies, Paul

God and The New Physics
Touchstone Books
Simon and Schuster Inc. N.Y.

Dillard, James M.D. D.C.
Ziporyn, Terry PhD

Alternative Medicine
for Dummies (1998)
I.D.G. Books Worldwide Inc.

Dossey, Larry M.D

Space, Time and Medicine
Shambhala (1982)
Boulder and London

Dubos, Rene (1970)

Reason Awake
Columbia University Press, New York

Durant, Will (1944)

The Story of Civilization
Simon and Schuster

Eisenberg, David & Kauzman, Walter

The Structure and Properties
of Water (1969)
Oxford University Press, New York and Oxford
Encyclopedia Brittania (1972 Edition)

Erikson, Eric (1963)

Childhood and Society
(Second Edition)
W.W.Norton and Company Inc.

Forgus, Ronald (1966)

Perception
McGraw-Hill

Freud, Sigmund (1959)

Collected Papers (5 volumes)
Translation- Joan Riviere
Basic Books

Fundenberg, H.H. et al.(1976)

Basic and Chemical
Immunology (Second Edition)
Lange Medical Publications

Gach, Michael Reed

Acupressure Potent Points (1990)
Bantam Books N.Y.

Galland, Leo M.D.   Power Healing (1997)
                                             Random House, N.Y.

Gamow, George (1963)   The Atom and the Nucleus
                                             Prentice Hall Inc.

Gaskell, Walter H. (1920)   The Involuntary Nervous System
                                      Longmans,Green and Co, London

Gesell, Arnold and
Ilg, Francis L (1949)   Child Development
                                      Harper and Row Publishers, New York

Goodman, Louis & Gilman, Alfred (!980)   Pharmacologic Basis of
                                      Therapeutics (Sixth Edition)
                                      Macmillon Publishing Co, Inc.

Glasser, William (1965)   Reality Therapy
                                      Harper and Row Publishers, New York

Glasstone, Samuel (1967)   Sourcebook of Atomic Energy
                                             (Third Edition)
                                      Van Nostrand Reinhhold Co.

Hall, Edward (1977)   Beyond Culture
                                      Doubleday Anchor Books

Hawking, Stephen (1988)   A Brief History of Time
                                             Bantam Books

Hawkins, David R. M.D. PhD.   Power vs. Force (2002)
                                             Hay House Inc.

Hayek, F.A. (1988)   The Fatal Conceit
                                      University of Chicago Press

Hayek, F.A. (1960)   The Constitution of Liberty
                                      University of Chicago Press

Hunt, Morton (1982)   The Universe Within
                                             Simon and Schuster

Illingsworth, R.S. (1972)
The Development of the Infant
and Young Child
Williams and Wilkins Co.

Jasper, Karl (1963)
General Psychopathology
Translated from German by J. Hoenig and Marian Hamilton
University of Chicago Press

Jones, D&K. (1960)
Structural Psychology
Pergamon Press Inc, New York

Kass, Leon (1985)
Toward a More Natural Science
The Free Press, Macmillon Inc.

Kiell, Norman
The Universal Experiences of Adolescence
Beacon Press, Boston

Kinsey, Alfred et al. (1953)
Sexual Behavior
of the Human Female
W.B. Saunders Company
Philadelphia, London

Kinsey, Alfred et al. (1948)
Sexual Behavior
of the Human Male
W.B. Saunders Company
Philadelphia, London

Kissinger, Henry (1994)
Diplomacy
Simon and Schuster

Koestler, Arthur (1973)
The Act of Creation
Dell, Publishing Co, Inc.

Korzybski, Alfred (1941)
Science and Sanity
(Second Edition)
The Internation Non-Aristotelian
Library Publishing Company

Kraft Ebing, Richard Von
Psychopathia Sexualis
Pioneer Publications Inc.

Lawrence, Ron M.D. PhD (1998)   Magnet Therapy - The Pain Cure
Rosch, Paul J. M.D. FACP
Prima Publishing

Leukel, Francis (1968)   Introduction to
Physiological Chemistry
C.V. Mosby Company

Levinson, Daniel   The Seasons of a Man's Life (1995)
Alfred A. Knopf, N.Y.

Lieberman, Philip (1984)   The Biology and Evolution of Language
Harvard University Press
London, England

Lown, Bernard M.D.   The Lost Art of Healing (1996)
Houghton Mifflin Co.

Mahler, Margaret et al. (1975)   The Psychologic Birth
of The Human Infant
Basic Books Inc, New York

Masters,William & Johnson,Virginia   Human Sexual Response (1966)
Little Brown and Company, Boston

Matsumoto, Tero M.D. PhD. FACS   Acupuncture For Physicians (1974)
Charles Thomas Publisher
Springfield, Ill.

Micozzi, Marc S.   Fundamentals of
Complimentary and Alternative Medicine
Churchill Livingstone (1996)

Miles, Stanley   Underwater Medicine (1962)
J.B. Lippincott Co.
Philadelphia, PA.

Morris, Richard (1984)   Dismantling the Universe
Touchstone Book- Simon and Schuster

Moyers, Bill

Healing and The Mind (1995)
Doubleday Inc.

Murphy, Lois Barclay & Moriarity, Alice

Vulnerability, Coping
and Growth (1976)
New Haven and London Yale University Press

Neibuhr, Reinhold (1937)

Beyond Tragedy
Charles Scribner's Sons

Neibuhr, Reinhold (1955)

The Self and the
Dramas of History
Charles Scribner's Sons

Neibuhr, Reinhold (1949)

The Nature and Destiny of Man
Charles Scribner's Sons

Cognitive and Mental Development
National Institute of Mental Health

Moral Man and Immoral Society
Charles Scribner Sons, New York

Nietzsche, Friedrich (1967)

Thus Spoke Zarathustra
Heritage Press

Opik, Ernet (1960)

The Oscillating Universe
A Mentor Book-New American Library

Ornstein, Robert (1973)

The Nature of Human Consciousness
The Viking Press, New York

Ornstein, Robert (1987)

The Healing Brain
Simon and Schuster

Osler, Sir William (1947)

Aequanimitas
Blakiston Co., Philadelphia

Pauling, Linus (1980)

The Nature of the Chemical Bond
(Third Edition)
Cornell University Press

Wallace Salzman

Pert, Candace (1997)   Molecules of Emotion
Charles Scribner, New York

Piaget, Jean (1952)   The Origin of Intelligence in Children
International Universities Press, New York

Piaget, Jean (1969)   The Psychology of the Child
Basic Books: A Harper Torch Book

Piaget, Jean (1954)   The Construction of Reality in The Child
translated by Margaret Cook
Basic Books Inc, New York

Rand, Ayn (1957)   Atlas Shrugged
New American Library

Rosenfelt, Isaac et al. (1983)   DNA for Beginners
Writers and Readers Cooperative Ltd. London

Rudyar, Dane (1970)   The Astrology of Personality
Doubleday Paperback

Sagan, Carl (1980)   Cosmos
Random House, New York

Salzman, Wallace (1951 to 1999)   Post Graduate Courses at
Hoektoen Institute- Chicago
Stanford University- Palo Alto
Harvard University- Boston
Unpublished Notes-1951-2001 (30 Volumes)
Ortho-Para I, II, III

Santayana, George (1954)   The Life of Reason
Charles Scribner Sons, New York

Siegman, A.E. (1971)   Introduction to Lasers and Masers
McGraw-Hill Book Company

Siegal, Bernie M.D.   Love, Medicine and Miracles (1986)
Harper and Row

Smoot, Robert et al. (1983)

Chemistry
Charles Merrill Publishing Co.

Skinner, B.F. (1953)

Science and Human Behavior
The Free Press, New York

Spitz, Rene (1957)

No and Yes
International Universities Press Inc.

The First Year of Life
International Universities Press (1965)

Smith, Homer (1961)

From Fish to Philosopher
Natural History Library-Anchor Books
Doubleday and Company

Schatzberg, Alan & Nemeroff, Charles

Psychopharmacology (1995)
American Psychiatric Press Incorporated

Speight, Robert (1967)

The Life of Teilhard de Chardin
Harper and Row Publishers, New York

Wallerstein, Judith et al.

The Unexpected Legacy of Divorce
Hyperion, New York

Washnic, George J
Hricak, Richard Z.

Discovery of Magnetic Health (1993)
Nova Publishing Co.
Rockville, Maryland

Weil, Andrew M.D.

Spontaneous Healing (1995)
Ballantine Books, N.Y.

Wiener, Norbert (1965)
Schade, J.P.

Progress in Biocybernetics
Elsevier Publishers

Weinberg, Steven (1977)

The First Three Minutes
Basic Book Inc. Publishers

Wilber, Ken (1983)

Eye to Eye
Anchor Press/Doubleday

Williams, Harley
The Healing Touch (1952)
Charles C. Thomas Publisher
Springfield, Il.

Wilson, Edward O.
Consilience: The Unity Of Knowledge
Vintage Books Inc, New York

Yankelovitch, David (1970)
Barret, William
Ego and Instinct
Random House, New York

Young, Stewart (1984)
NMR Imaging
Raven Press, New York

Index

www.ingramcontent.com/pod-product-compliance
Lightning Source LLC
Chambersburg PA
CBHW031821170526
5157CB00001B/139